RATIONALITY IN SCIENCE

RATIONALITY IN SCIENCE

Studies in the Foundations of Science and Ethics

Edited by

RISTO HILPINEN

University of Turku, Turku, Finland

D. REIDEL PUBLISHING COMPANY

DORDRECHT : HOLLAND / BOSTON : U.S.A.
LONDON : ENGLAND

Library of Congress Cataloging in Publication Data

Main entry under title:

Rationality in science.

    (Philosophical studies series in philosophy ; v. 21)

    Papers from a conference held in Dubrovnik, Yugoslavia, Mar. 6–12, 1978, sponsored by the Inter-university Centre of Post-graduate Studies, Dubrovnik.

    Includes bibliographical references and indexes.

    1. Rationalism—Congresses. 2. Science—Philosophy—Congresses. 3. Empiricism—Congresses. 4. Ethics—Congresses. I. Hilpinen, Risto. II. Inter-university Centre of Post-graduate Studies, Dubrovnik, Yugoslavia.

BD181.R36   121   80–18853

ISBN 90-277-1112-7

Published by D. Reidel Publishing Company,
P.O. Box 17, 3300 AA Dordrecht, Holland.

Sold and distributed in the U.S.A. and Canada
by Kluwer Boston Inc.,
190 Old Derby Street, Hingham, MA 02043, U.S.A.

In all other countries, sold and distributed
by Kluwer Academic Publishers Group,
P.O. Box 322, 3300 AH Dordrecht, Holland

D. Reidel Publishing Company is a member of the Kluwer Group

CONTENTS

PREFACE vii

LARS BERGSTRÖM / Some Remarks Concerning Rationality in
Science 1

RISTO HILPINEN / Scientific Rationality and the Ethics of Belief 13

KAREL LAMBERT / Explanation and Understanding: An Open
Question? 29

ADRIENNE LEHRER / The Empirical Investigation of Synonymy
and the Implication for Science 35

KEITH LEHRER / A Model of Rational Consensus in Science 51

KUNO LORENZ / Science, a Rational Enterprise? Some Remarks
on the Consequences of Distinguishing Science as a Way of
Presentation and Science as a Way of Research 63

MIHAILO MARKOVIĆ / Scientific and Ethical Rationality 79

WILLIAM NEWTON-SMITH / The Underdetermination of
Theory by Data 91

ROLAND POSNER / Types of Dialogue – The Use of Microstructures for the Classification of Texts 111

MARIAN PRZEŁĘCKI / Conceptual Continuity through Theory
Changes 137

IVAN SUPEK / Science and Humanism 151

PATRICK SUPPES / Probabilistic Empiricism and Rationality 171

KNUT ERIK TRANØY / Norms of Inquiry: Rationality, Consistency Requirements and Normative Conflict 191

JULES VUILLEMIN / The Influence of Reason on the Origin of
Science 203

PAUL WEINGARTNER / Normative Characteristics of Scientific
Activity 209

DAGFINN FØLLESDAL / Explanation of Action 231

INDEX OF NAMES 249

INDEX OF SUBJECTS 253

PREFACE

The present volume is a product of an international research program 'Foundations of Science and Ethics', launched in 1976 by the Inter-University Centre of Post-Graduate Studies, Dubrovnik, Yugoslavia, with the financial support of the Volkswagen Foundation. According to the outline of the program, formulated in 1976 by a committee consisting of Professors Dagfinn Føllesdal, Rudolf Haller (coordinator), Lorenz Krüger, Karel Lambert, Keith Lehrer, Kuno Lorenz, Günther Patzig, Ivan Supek and Paul Weingartner, its general purpose was to investigate the interplay of various internal and external factors in the development of science. Generous financial support from the Volkswagen Foundation made it possible to plan four annual conferences, the first of which was held in Dubrovnik on March 6–12, 1978. This volume contains the majority of the papers presented in the first Dubrovnik conference; the main theme of this conference was 'Rationality in Science and Ethics' (Some of the papers appear here in a thoroughly revised form.) Further results of the research program will be discussed in three other conferences, to be held in Dubrovnik in 1979–1981; the papers presented in these conferences will be published separately.

Professor Rudolf Haller of the University of Graz assumed the burden of the practical planning and organization of the first conference (as well as that of the other three conferences). I wish to thank Professor Haller on behalf of all participants for carrying out this demanding and time-consuming task. My thanks are also due to the Volkswagen Foundation whose financial support made the conferences possible, and to the Inter-University Centre of Dubrovnik for providing a pleasant and inspiring atmosphere for our discussions.

THE EDITOR

LARS BERGSTRÖM

SOME REMARKS CONCERNING
RATIONALITY IN SCIENCE

Someone might argue as follows: "Man is a rational animal, and his rationality is surely most prominent in science. No other major area within human culture can compete with science as far as rationality is concerned. Hence, by studying the place and nature of rationality in science, one may learn something important about rationality in general. This would be useful in many other contexts. Besides, a study of rationality in science might make it possible for us to improve the functioning and organization of scientific activities. Perhaps science can thereby be made even more rational than it already is."

Personally, I find it hard to believe that scientific activities are more rational than other activities, or that scientists are more rational than people in general. Neither is it clear to me that rationality is a desirable feature in scientific activities, or that science ought to be made more rational than it already is, or that it ought to be made as rational as possible. For all I know, each of these claims *may* indeed be acceptable, but whether or not they *are* acceptable seems to me to be an open question. Part of the problem here is that the very notion of 'rationality' can be interpreted in various ways. In this paper I shall try to indicate a number of different interpretations all of which seem comparatively reasonable from the point of view of ordinary language. I shall also make some suggestions as to the possibility and desirability of different kinds of rationality in science.

1. Methodological Rules

Paul Feyerabend claims that "the attempt to make science more 'rational' and more precise is bound to wipe it out".[1] In other words, he seems to hold that rationality is a threat to scientific progress. Some people may find this view rather surprising. But what is meant by 'rationality' here? It seems that what Feyerabend is saying is that scientific activity is not governed – and should not be governed – by methodological rules. For example, he says that

one of the most striking features of recent discussions in the history and philosophy of

1

R. Hilpinen (Ed.), Rationality in Science. 1–11.

science is the realization that events and developments, such as the invention of atomism in antiquity, the Copernican Revolution, the rise of modern atomism (kinetic theory; dispersion theory; stereochemistry; quantum theory), the gradual emergence of the wave theory of light, occurred only because some thinkers either *decided* not to be bound by certain 'obvious' methodological rules, or because they *unwittingly broke* them.[2]

The methodological rules which Feyerabend has in mind here are primarily those which have been proposed by critical rationalists (e.g. Popper) and logical empiricists:

wherever we look, whatever examples we consider, we see that the principles of critical rationalism (take falsifications seriously; increase content; avoid *ad hoc* hypotheses; 'be honest' – whatever *that* means; and so on) and, *a fortiori*, the principles of logical empiricism (be precise; base your theories on measurement; avoid vague and unstable ideas; and so on) give an inadequate account of the past development of science and are liable to hinder science in the future.[3]

Feyerabend is not saying that it would be irrational for scientists to act in accordance with methodological rules of the kind indicated here. On the contrary, he seems to hold that acting in accordance with such rules is precisely what is meant by 'rationality' in science. He might even agree with Popper that "there is nothing more 'rational' than . . . the method of science".[4] But he holds that this kind of rationality is undesirable and not very common: the method of science would be an obstacle to scientific progress.

I do not know to what extent Feyerabend is right in his interpretation of the scientific development so far, but I would not be surprised if he is largely right. However, even if scientists have often behaved irrationally in the past, and even if several important theories would not have emerged when they did if their authors had behaved more rationally, surely it does not follow that rationality would also hinder scientific progress in the future.

Nevertheless, I am inclined to believe that 'rationality' in the sense discussed here is probably not very important or desirable in science. The reason is that the methodological rules in question seem to confuse means and ends, or process and product, in a certain way. It seems that these rules are plausible in so far as they merely indicate certain desirable features of the ultimate products of scientific research. It is certainly desirable – other things being equal – that those theories which represent the ideal outcome of scientific research are not falsified, that they have a comparatively large content, that they are not *ad hoc*, that they are formulated in a precise way, and so on. In other words, the so-called methodological rules may plausibly be understood as stating (some of) the *aims* of science. They need *not* be taken to prescribe any particular behavior on the part of individual scientists. In particular, they need not be taken to prescribe the immediate abandonment of theories or hypothesis which are falsified, *ad hoc*, vague, unstable, and so on. Rationality may not always be a good means to the attainment of the aims of science.

2. UTILITY-MAXIMIZATION

In economics, decision-theory and game-theory rationality is often identified with some kind of utility-maximization. A rational agent acts in such a way that his wants or preferences will (probably) be satisfied to a maximal degree. For example, Luce and Raiffa write as follows:

Though it is not apparent from some writings, the term "rational" is far from precise, and it certainly means different things in the different theories that have been developed. Loosely, it seems to include any assumption one makes about the players maximizing something, and any about complete knowledge on the part of the player in a very complex situation, where experience indicates that a human being would be far more restricted in his perceptions.[5]

Similarly, according to John Rawls

a rational person is thought to have a coherent set of preferences between the options open to him. He ranks these options according to how well they further his purposes; he follows the plan which will satisfy more of his desires rather than less, and which has the greater chance of being successfully executed.[6]

Rawls claims that this is the standard concept of rationality in social theory.

I take it that rationality in this sense involves knowledge. This is often made quite explicit. One would not say that a person is rational – or even that his activity is rational – if he just *happens* to maximize his utility. A rational agent (in this sense) knows what his alternatives are, and he knows what the possible outcomes of these alternatives are. He also knows what his preferences are among the possible outcomes. In short, he *knows* what alternatives will maximize his utility, and he chooses one of these alternatives *because* it will maximize his utility (or his expected utility).

There seems to be no reason at all to believe that scientists are generally rational – or more rational than other people – in this sense.

Scientific activities certainly involve a number of more or less conscious choices. For example, scientists choose problems, collaborators, methods, key concepts, definitions, hypotheses, criteria of sufficient evidence, and so on. But it seems plausible to assume that scientists are seldom if ever aware of all the alternatives open to them and all the possible outcomes of these alternatives. This is not only because our knowledge of such things is always limited in real life. Apart from this, I would suggest that our ability to predict the outcomes of various research strategies is especially limited. It is an important characteristic of scientific activities that they may lead to new discoveries which cannot be known in advance, and these may in turn have consequences which are even more difficult to predict. As far as I can see, this is the main reason why scientists are not – and cannot be expected to be – rational in the sense indicated here.

3. SUBJECTIVE UTILITY-MAXIMIZATION

At this point it may seem natural to give up the knowledge requirements and to say instead that a person – or his actions or his activity in a given situation – is rational if and only if he acts in such a way that, from the point of view of those beliefs (and desires) that he actually has at the time of acting, his actions can be expected to maximize his utility. This seems to be the notion that David Gauthier has in mind when he writes that

the rational man acts to bring about that outcome which he prefers, among those which he *believes* are open to him.[7]

Hempel has formulated a similar proposal as follows:

To qualify a given action as rational is to put forward an *empirical hypothesis* and a *critical appraisal*. The hypothesis is to the effect that the action was done for certain reasons, that it can be *explained* as having been motivated by them. The reasons will include the ends that the agent presumably sought to attain, and the beliefs he presumably entertained concerning the availability, propriety, and probable effectiveness of alternative means of attaining those ends. The critical appraisal implied by the attribution of rationality is to the effect that, judged in the light of the agent's beliefs, the action he decided upon constituted a *reasonable* or *appropriate* choice of means for achieving his end.[8]

Rationality in this sense is obviously a relative concept. Whether a given action – or the decision to perform it – is rational will depend on the objectives the action is meant to achieve and on the relevant empirical information available at the time of the decision. Broadly speaking, an action will qualify if, on the given information, it offers optimal prospects of achieving its objectives.[9]

When Hempel speaks here about "the relevant empirical information available at the time of the decision", I take it that he is referring to those beliefs which the agent actually entertains at the time of the decision and which actually motivates his decision (according to our empirical hypothesis). Information which is only "available" to the agent in some other sense is presumably not relevant to the question of whether or not his action is rational in Hempel's sense. Moreover, it seems that Hempel would say that an action "offers optimal prospects of achieving its objectives" if it maximizes the agent's utility or his expected utility.[10] In the case of decisions under uncertainty he mentions the maximin rule ("maximize minimum utility") and the maximax rule ("maximize maximum utility") as possible criteria of rational choice.[11] For simplicity, I shall here use the phrase "maximizes the agent's utility" to cover all these alternatives (or the relevant ones). It seems that we may then summarize Hempel's proposal as follows:

(H)   A particular action a is rational if and only if the agent's effective reasons for doing a constitute good reasons for believing that a maximizes the agent's utility.

Now suppose that an agent performs a certain action, and that among his effective reasons for performing the action is the belief that the action maximizes his utility. Does it follow that the action is rational in Hempel's sense? I think not. That is why I have chosen the formulation (H), which makes use of the notion of "good reasons". For I am assuming that, in general, the belief that p does not constitute a good reason for believing that p. However, I may be wrong about Hempel's intentions with respect to this point. Perhaps he would say that a person acts rationally if he believes that the action he performs will maximize his utility. In that case, the following formulation may be better suited to Hempel's intentions:

(H')   A particular action a is rational if and only if the agent's effective reasons for doing a include the belief that a maximizes the agent's utility, or constitute good reasons for this belief.

Are scientific activities rational in Hempel's sense? I am inclined to think that they are seldom rational according to (H). Even if scientists often have fairly good reasons for *doing* what they do, these reasons would probably seldom be good reasons for *believing* that the actions in question will maximize their utilities. This may sound paradoxical, but I think it can be explained by a simple example. Suppose that X believes that he must choose one of the two alternatives a_1 and a_2, and that he wants to maximize his utility. Suppose further that X's reasons for believing that a_1 will maximize his utility are somewhat stronger than his reasons for believing that a_2 will maximize his utility. In that case it seems that X has a good reason for *doing a_1*. But it does not follow that he has thereby also a good reason for *believing* that a_1 will maximize his utility (or his expected utility). Given his evidence, it may be more reasonable to suspend judgment. In general, as long as a person has good reasons for doing what he does, he does not *need* to have, in addition, good reasons for believing that what he does will maximize his utility. (Another thing is, of course, that his position is improved in a certain way if he also has good reasons of the second type.)

If this distinction is accepted, I think it will also be agreed that there is really no reason to believe that scientists generally have good reasons to believe that their activities will maximize their utilities. On the contrary, everything seems to point in the direction of the opposite assumption. I have already commented upon this, and Hempel would presumably agree. He says that the usual decision-theoretical models of rational behavior.

do not offer us much help for a rational solution of the grave and complex decision problems that confront us in our daily affairs. For in these cases, we are usually far from having the data required by our models: we often have no clear idea of the available courses of action, nor can we specify the possible outcomes, let alone their probabilities and utilities.[11]

In the case of parlor games the relevant information may often be at least theoretically available, but Hempel's characterization is presumably applicable to most scientific contexts. Hence, in general we cannot expect scientific activities to be rational in the sense of (H).

On the other hand, scientific activities may fairly often be rational in the sense of (H'). Personally, I do not think they are, but it is at least possible. Scientists may of course believe that their activities will maximize their utilities. But, for the reasons indicated above, such beliefs would be irrational in most cases, i.e., they would not be based upon good reasons. Hence, although rationality in the sense of (H') is possible, it does not seem very desirable in science.

4. GOOD REASONS FOR ACTING

I have indicated above that a scientist may sometimes have good reasons for doing what he does. Perhaps this is all that some people mean by "rationality". For example, Quentin Gibson says that

It is only when someone has a *good* reason for acting that we speak of him as acting *rationally*.[12]

A few lines later he says that

people sometimes act rationally - in the sense of having good reasons for their actions.[12]

Of course, one may wonder what is meant here by "good reasons" for acting. It might be held, for example, that a person has a good reason for doing a certain action if and only if he believes that the action will maximize his utility. In that case we are back to (H'), or something very close to (H'). But Gibson explains the notion of "good reasons" for acting as follows:

To have a good reason for acting, three things are necessary. There must be a belief that the action will further one's ends. There must be sufficient evidence available to one to make this belief a reasonable one to hold. And finally one must *take account* of this evidence and appreciate that it is sufficient to justify the belief.[13]

Presumably, Gibson means that these three factors are not only necessary, but also jointly sufficient, for someone to have a good reason for acting. On the other hand, I am not sure whether he really means that having a good reason in this sense is sufficient for rationality, or whether he would add - in the same way as Hempel - that rationality also presupposes that the good reasons are *effective* reasons, or that the action in question is *motivated* by the agent's belief that it will further his ends and his appreciation that the available evidence is sufficient to justify this belief.

However, Gibson's proposal seems to be different from those expressed by (H) and (H') in any case. The main difference - or the difference that I want to emphasize here - is that Gibson's criterion of rationality does not involve any notion of maximization. It seems that a person may act rationally in Gibson's sense without knowing or believing anything at all about any alternatives to his action. It is sufficient that he believes, on good grounds, that the action in question will further his ends. (It may be noticed that this is quite in accordance with my

suggestion above that one may often have good reasons for doing something without having good reasons for believing that it will maximize
one's utility. On the other hand, Gibson's explication of the notion of
having a good reason for acting is not consistent with the example
which I used to illustrate this possibility.)

Are scientific activities rational in Gibson's sense? In other words, do
scientists generally believe, on good grounds, that their activities will
further their ends? It seems fairly plausible to assume that they do.
However, this is a rather weak sense of rationality, so that even if it is
desirable that scientific activities are rational in this sense, it does not
seem desirable that they are rational *merely* in this sense. For example, it
also seems desirable that scientists are rational in some sense which
implies that they take at least some account of various alternatives
which may confront them.

5. REASONABLE AIMS

There is one obvious reason for doubting or denying the desirability of
rationality in Gibson's sense, or in the senses represented by (H) and
(H'), namely that the aims and preferences that individual scientists
have may be such that it would be better if they were not satisfied.

There is a great variety of aims that may motivate scientific activities.
According to the outline for this research project, the "most general aim
guiding the scientific enterprise is to obtain empirical truth and theoretical understanding". Moreover, it is claimed that "simplicity, consistency, coherence, and comprehensiveness are other general and perhaps
invariant goals of contemporary science". These aims may be called
intra-scientific. Of course, intra-scientific aims may usually be much
more specific, as when a scientist wants to find the answer to a particular
question. However, it can be safely assumed that scientists often have
other aims (as well) which are of a more extra-scientific nature, but
which may nevertheless influence their scientific activities. For example,
some scientists want to become famous and most want to be successful,
some want to be invited to congresses and to travel around the world,
some want to get rich and to be awarded academic honours, some want
to refute ideas which are proposed by people they dislike and to silence
their own critics, some want to contribute to the well-being of others,
some want to support certain political ideals, some want their work to
have certain aesthetic qualities, and so on.

It seems that extra-scientific aims are particularly influential when scientists choose their problems or topics. From a methodological point of view such choices may be rather peripheral, but they seem to be very important as determinants of scientific development.

Now it might be held that for a scientist to be engaged in a rational activity it is not enough that he believes, on good grounds, that the activity in question will further his ends, but that it is also necessary that his ends are reasonable in some sense.

What could be meant here by the requirement that his ends be reasonable? Let me make the following suggestion. We may say that an end E is reasonable for an agent X if and only if X wants to attain E and there are no propositions $p_1, \ldots, p_n$ such that there are good reasons, which are available to X, for believing $p_1, \ldots, p_n$, and if X were to believe $p_1, \ldots, p_n$ he would not want to attain E. Of course, this is only a rather rough characterization. It could be modified and refined in various ways, but I shall not go into that here.

It should be noticed, however, that even if a scientist's aims or ends are reasonable in this sense it may not be desirable, or even tolerable, that he attains, or tries to attain, them. In such cases, it is not desirable that he acts rationally in the sense indicated above.

6. WELL-PREPARED CHOICES

A person may maximize his utility or attain his ends more or less by chance. In that case we might want to say that he is lucky rather than rational. It seems that in one sense of the word 'rational' we would say that an activity is rational only if it involves, or is based upon, a fully conscious and comparatively elaborate process of deliberation. We might even be inclined to say that an agent, and his activity, is rational if he goes through such a process – regardless of whether the results of this process, i.e., his choices, actually turn out to be rational in any of the senses mentioned above.

A reasonable and thorough process of deliberation would presumably include the following elements: (i) a serious attempt to formulate in some detail the main goals of the activity to be chosen, (ii) a surveying of available information which may be relevant to the choice of means of attaining these goals, (iii) a fairly detailed structuring of the choice-situation, involving a list of possible alternatives and a list of possible total outcomes for each of these alternatives, (iv) an evaluation or preference-ordering of the possible outcomes, (v) a critical questioning, which may lead to modifications, of the results of (i) through (iv), (vi) a surveying of various criteria of choice, and a discussion and evaluation of these, (vii) an application of some preferred criterion of choice and a decision in accordance with this criterion.

A process of this kind may or may not lead to a belief that the chosen alternative will maximize the agent's utility, or to a belief, which is based upon sufficient evidence, that it will further the agent's ends. Whether or not the process will lead to such a belief depends, among other things, upon the available evidence, the agent's imagination, and his degree of scepticism. But regardless of whether it leads to such a belief, we might want to say that the agent's activity is rational if it is based upon, or involves, a deliberation process of this kind.

However, it might be objected that it would be irrational, in many cases, for an agent to engage in a deliberation process of this kind. For example, it would be irrational – and almost a sign of mental illness – for a person to engage in such a process every time he is to select a tie or get out of bed in the morning. A careful and detailed deliberation procedure involves great costs in terms of time and effort, and in certain cases these costs may be much too great. Hence, as an alternative to the proposal indicated above, it might be suggested that an activity is rational if and only if it involves, or is based upon, a process of deliberation which is just as extensive and careful as is fitting in a situation of the kind in question.

Personally, I am inclined to believe that scientific activities are seldom rational in the first of the two senses suggested here. Neither does it seem very desirable that they should be. On the other hand, scientific activities may often be rational in the second sense. Besides, it seems trivially true that this is desirable. However, the application of this second criterion of rationality presupposes a personal judgment or evaluation. Different people may easily come to different conclusions when they apply it to a given case, and it is hard to see how one should settle disagreement of this kind. Hence, this criterion may not be very useful in practice.

7. CONCLUDING REMARKS

I have argued that rationality is presumably not very common and not very desirable in science. At least, this seems to be true for most senses of the rather ambiguous term 'rationality', even if it is not true for all senses.

I have been exclusively concerned with rationality as an attribute of persons and of their activities and decisions. But in the present research project it seems that rationality is thought of as an attribute of scientific development. It is not at all clear to me what could be meant by this. However, any scientific development certainly involves activities on the part of individual scientists. Hence, it might be held that a certain scientific development is rational to the extent that the activities which it involves are rational. With this interpretation my discussion above seems to be relevant. But perhaps other interpretations are possible. For example, it might be suggested that a scientific development is rational to the extent that it has been planned and to the extent that it proceeds according to this plan. It seems plausible to assume that scientific developments are seldom if ever rational in this sense. It is also very doubtful whether they *can* be rational in this sense. Can they be rational in any other sense? This seems also rather doubtful.

Uppsala University

NOTES

[1] Paul Feyerabend, *Against Method*, London 1975, p. 179.

[2] *Ibid.*, p. 23.

[3] *Ibid.*, p. 179.

[4] Karl Popper, *Objective Knowledge*, London 1972, p. 27.

[5] R. Duncan Luce and Howard Raiffa, *Games and Decisions*, New York 1957, p. 5.

[6] John Rawls, *A Theory of Justice*, Oxford 1973, p. 143.

[7] David Gauthier, 'Reason and Maximization', *Canadian Journal of Philosophy*, 4 (1975), p. 415, my italics.

[8] Carl G. Hempel, *Aspects of Scientific Explanation*, New York 1966, p. 463.

[9] *Ibid.*, p. 464.

[10] *Ibid.*, p. 466.

[11] *Ibid.*, p. 467.

[12] Quentin Gibson, *The Logic of Social Enquiry*, London 1960, p. 43.

[13] *Ibid.*, pp. 43–44.

RISTO HILPINEN

SCIENTIFIC RATIONALITY
AND THE ETHICS OF BELIEF

I

It may seem paradoxical to apply normative expressions, such as "ought", "may", "right", etc., to a person's beliefs. Normative concepts are primarily applicable to acts, and their main function is to regulate conduct, but beliefs (or believings) are not acts: believing something is not an act or an event, but rather a *state* of an individual. Moreover, a person's actions are subject to ethical evaluation and criticism only so far as they are free: if a person could not have committed a certain act *F*, it is pointless or nonsensical to say that he or she ought to have done it. But we cannot normally choose our beliefs: if a person believes something, he cannot simply decide to withhold his belief, or disbelieve what he previously believed. Many beliefs, e.g. perceptual beliefs, are almost compulsory: for example, if a person clearly sees that a certain object in front of him is red, he cannot help believing that the object is red.

In spite of these apparent differences between believing and free action, it is often perfectly natural to tell a person that he or she ought or ought not to believe something, or has no right to believe something. Perhaps beliefs cannot be chosen at will, but people nevertheless seem to have, or are assumed to have, some freedom and control over what they believe.

The question of the relationship between belief and will has been subject to philosophical dispute and controversy for centuries. Philosophers have held sharply divergent views on this issue: for example, in his *Meditations* Descartes assumes that belief or "judgment" is an act of free will, whereas Hume maintained that believing (or assent) is wholly independent of the will.[1] The truth of the matter seems to lie somewhere between these extreme positions. According to P. T. Geach,

Beliefs cannot be immediately switched on or off at will, but they are to some extent under our control. We can form habits of thought which will modify our beliefs for good or ill, and the formation of such habits is certainly voluntary.[2]

As was mentioned above, beliefs are not actions or events, but a *change* of belief is an event. A person can learn to change his beliefs on the basis

R. Hilpinen (Ed.), Rationality in Science. 13–28.
Copyright © 1980 by D. Reidel Publishing Company.

of inference and reasoning in a deliberate manner; in such cases his beliefs do not only change, but he changes his beliefs, and they can plausibly be regarded as results of free actions. Normative statements concerning beliefs resemble *directives*, i.e., normative utterances in which a person is advised or commanded to do (or refrain from doing) something. Thus the statement "*a* ought to believe *h*", considered as a directive, may be taken to mean that *a* ought to change his belief system so as to include *h* in it (if *h* is not already included), or that *a* ought to continue to believe *h* (if he already believes *h*). Belief change is a complex process, and even though a person cannot change his beliefs by simple acts of will, he can undertake to change them by various means, e.g. by inquiry and deliberation.

II

In his paper 'Truth, Fallibility, and the Growth of Knowledge'[3] Isaac Levi distinguishes four types of belief change. Let B_0 be a person's (or an agent's) original system of beliefs, and let B_1 be the system resulting from the change. According to Levi, the shift from B_0 to B_1 may belong to one of the following types:

(i)   Expansion: B_0 is a proper subset of B_1.
(ii)   Contraction: B_1 is a proper subset of B_0.
(iii)   Replacement: B_1 is obtained from B_0 by replacing some $h \in B_0$ by its negation.
(iv)   Residual shift: any shift which is not of type (i)–(iii).

It is clear that any replacement can be defined as a contraction and a subsequent expansion; thus replacements can be analysed in terms of simpler changes which belong to types (i) and (ii). In this paper I shall make some elementary remarks on the normative principles of belief expansion. (The ethics of belief contraction is a more difficult topic, and I shall not attempt to take it up here.)

According to Levi, a belief expansion can be (a) an inferential expansion or (b) a routine expansion. Inferential expansions are based on inquiry and deliberation, and they possess all the main characteristics of free action. In routine expansions the agent lets some stochastic process determine what he adds to his system of beliefs. For example, if *a* adds new propositions to B_0 as a direct response to sensory stimulation, or asks another person, b, whether h is true, and lets b's reply determine whether he adds h to B_0, he is changing his beliefs by routine expansion.[4] Both routine expansions and inferential expansions are subject to rational evaluation and criticism, but in the case of the former the criticism concerns the expansion procedures, not the individual propositions added to B_0: when a adds new propositions to his belief system by routine expansion, the probability that the expansion procedure will introduce false propositions into the system should be small. In the following I shall discuss only deliberate or inferential expansions, in which we may pass a judgment on each individual expansion "act" separately.

III

If we consider belief on a fairly rough level of analysis, and disregard all questions concerning the *degrees* of belief, it is possible to distinguish three different belief-attitudes or *doxastic attitudes* (of an agent a) towards a given proposition h:

A_1: a believes h; briefly: $B_a h$.

A_2: a neither believes nor disbelieves h (a withholds h or suspends judgment with respect to h); briefly:

$$W_a h \equiv \sim B_a h \mathbin{\&} \sim B_a \sim h.$$

A_3: a disbelieves h; briefly: $B_a \sim h$.

The alternatives A_1–A_3 are mutually exclusive and jointly exhaustive; thus *not believing* h is equivalent to the disjunction of A_2 and A_3, and *not believing* $\sim h$ to that of A_1 and A_2.[5] Suppose that a proposition h is subject to deliberation, and a has to decide which doxastic attitude towards h he ought to adopt. At the time of the deliberation his attitude towards h is of type A_2, and he can (i) expand his belief system by adding h to it, (ii) preserve the *status quo*, or (iii) expand his system by including $\sim h$ in it. Using the somewhat artificial terminology employed e.g. by Roderick M. Chisholm[6] and the Scandinavian action theorists, we may say that in case (i) a brings it about that he believes that h (or a sees to it that he believes that h), in (iii) a brings it about that he disbelieves h, and in (ii) he does neither (i) nor (iii). If a's bringing it about that A_i (or seeing to it that A_i) is expressed in terms of the

Do-operator of Kanger, Åqvist, Pörn and other Scandinavian action theorists,[7] (i)–(iii) can be formulated as follows:

(i) $Do_a B_a h$,

(ii) $\sim Do_a B_a h \ \& \ \sim Do_a B_a \sim h$,   and

(iii) $Do_a B_a \sim h$.

We are here concerned with a situation in which the agent is deliberately choosing a certain doxastic attitude towards h. A deliberate choice of (ii) can also be described as a's bringing it about that he neither believes nor disbelieves h; consequently (ii) can here be regarded as equivalent to

(ii') $Do_a(\sim B_a h \ \& \ \sim B_a \sim h)$.

The acts (i), (ii') and (iii) can be identified by their results ($B_a h$, $\sim B_a h \ \& \ \sim B_a \sim h$, and $B_a \sim h$, respectively), and in virtue of the equivalence of (ii) and (ii'), we may thus discuss the doxastic alternatives (i)–(iii) in terms of their results A_1–A_3.

If the doxastic alternatives A_1–A_3 (or (i)–(iii)) are considered from a normative standpoint, each of them may be regarded as obligatory (as one which ought to be chosen), as permitted (as one which may be chosen), or as forbidden (as one which may not be chosen); these deontic categories will be represented by the letters O, P, and F, respectively. The possibilities listed above are of course not independent of each other: I shall assume that O, P and F satisfy the standard principles of deontic logic and can be defined in terms of each other in the usual way. Thus the normative status of each doxastic alternative can be expressed in terms of a single deontic primitive, for example, in terms of P (the concept of permission or permissibility). I shall assume that P satisfies the standard laws[8]

(PI) $\sim P \sim A \supset PA$

and

(PII) $P(A \vee B) \equiv PA \vee PB$.

Each doxastic alternative A_i ($i = 1, 2, 3$) can be permitted or not permitted (for agent a in a given situation); thus we can define $2^3 = 8$ *basic normative positions* with respect to h as follows:[9]

$$P_1 = \quad PB_a h \ \& \quad PW_a h \ \& \quad PB_a \sim h$$
$$P_2 = \quad PB_a h \ \& \quad PW_a h \ \& \sim PB_a \sim h$$
$$P_3 = \quad PB_a h \ \& \sim PW_a h \ \& \quad PB_a \sim h$$
$$P_4 = \sim PB_a h \ \& \quad PW_a h \ \& \quad PB_a \sim h$$
$$P_5 = \quad PB_a h \ \& \sim PW_a h \ \& \sim PB_a \sim h$$
$$P_6 = \sim PB_a h \ \& \quad PW_a h \ \& \sim PB_a \sim h$$
$$P_7 = \sim PB_a h \ \& \sim PW_a h \ \& \quad PB_a \sim h$$
$$P_8 = \sim PB_a h \ \& \sim PW_a h \ \& \sim PB_a \sim h.$$

All conceivable normative positions with respect to h can be defined as disjunctions of the basic positions (or basic *conjunctions*) P_1-P_8; for example, $PB_a h$ is equivalent to the disjunction of P_1-P_3 and P_5, and $O \sim B_a \sim h$ is equivalent to $P_2 \vee P_5 \vee P_6 \vee P_8$. However, it is obvious that P_8 is inconsistent with (PI) and (PII): according to (PI), at least one of the alternatives A_i ($i = 1, 2, 3$) must be permitted. Thus P_8 may be struck out from the list of basic positions as a logically impossible case, but the remaining conjunctions P_1-P_7 are compatible with the principles of deontic logic, and if P_1-P_8 are regarded as abbreviations of more complex formulas involving the action alternatives (i)-(iii), they are also consistent with the logical principles governing the *Do*-operator.[10]

However, some of the basic positions P_1-P_7 conflict with our intuitions concerning permissible or reasonable belief. This conflict is especially clear in the case of P_3: P_3 describes a situation in which a may believe h and a may believe $\sim h$, but he may not withhold h (or suspend judgment with respect to h). In other words, in P_3 a may believe h or $\sim h$ (it does not matter which), but not withhold judgment. Such a situation seems logically unacceptable, and this suggests that the negation of P_3,

(1) $\quad \sim (PB_a h \ \& \sim P(\sim B_a h \ \& \sim B_a \sim h) \ \& \ PB_a \sim h)$,

that is,

(B1) $\quad PB_a h \supset (PW_a h \vee \sim PB_a \sim h)$,

is a valid principle of the ethics of belief. The action-theoretic analogue of (B1), viz.

(2) $\quad PDo_a A \supset (P(\sim Do_a A \ \& \sim Do_a \sim A) \vee \sim PDo_a \sim A)$

does not hold: it is possible that a may bring it about that A (or see to it that A) and a may see to it that $\sim A$, but he may not remain "passive" with respect to A and $\sim A$.[11] Thus (B1) cannot be derived from the general principles of deontic logic and action theory, but must be justified in terms of the special objectives and values which control deliberate belief expansion.

IV

It is often assumed that rational belief expansion is dependent on two main epistemic values or utilities, *truth* and *information*. William James has expressed this view as follows.[12]

> We must know the truth; and we must avoid error – these are our first and great commandments as would-be knowers; but they are not two ways of stating an identical commandment, they are two separable laws.

Several philosophers have recently developed theories of rational acceptance and belief on the basis of this Jamesian insight, and these theories are readily seen to vindicate principle (B1). I shall illustrate this in terms of the theory of rational belief expansion presented by Isaac Levi in his book *Gambling With Truth*.[13] According to Levi, the epistemic value or utility of adding h to one's belief system depends on two distinct factors, the *information content* of h and the *truth-value* of h. The Jamesian imperative "Believe truth!" reflects the epistemic utility of information (or content), and "Shun error!" the utility of truth.[14]

Levi assumes that epistemic utilities are measurable on an interval scale, and that the acceptability of a hypothesis depends on its expected epistemic utility.[15] If utilities are measurable on an interval scale, utility differences are additively measurable, i.e., unique up to the choice of the unit of measurement. Levi defines the difference between the epistemic utilities of two hypotheses as a linear combination of the corresponding differences in their truth-utilities and their content-utilities. Let $U_t(h)$ be the epistemic utility of accepting h if h is true, and $U_f(h)$ the epistemic utility of accepting a false hypothesis h. Let $T_t(h)$ and $T_f(h)$ be the truth-utilities ("truth-values") of a true and a false hypothesis h, respectively, and let $cont(h)$ be the content-value or informational utility of h. According to Levi,[16]

$$(3) \qquad U_x(h) - U_y(g) = \alpha(T_x(h) - T_y(g))$$
$$+ (1 - \alpha)(cont(h) - cont(g)),$$

where $x, y = t$ or f,

(4) $T_t(h) = 1$ and $T_f(h) = 0,$

and *cont* is a measure of the information content of a hypothesis, satisfying the following conditions:

(5) $0 \leqq cont(h) \leqq 1,$

(6) $cont(\sim h) = 1 - cont(h),$ and

(7) $cont(h \ \& \ g) = cont(h) + cont(g)$ if $h \vee g$ is a logical truth, i.e. if h and g have no common content.[17]

According to (4) and (5), *cont*-values and *T*-values are measured on similar scales; they have the same maximum and the same minimum value.

Let $\top$ be any logical truth (e.g. $h \vee \sim h$), and $\bot$ any logical falsehood (e.g. $h \ \& \sim h$). According to (5)–(7), $cont(\top) = 0$ and $cont(\bot) = 1$; thus

(8) $U_t(h) - U_t(\top) = (1 - \alpha)cont(h),$

(9) $U_f(\bot) - U_f(h) = (1 - \alpha)(1 - cont(h)),$

and

(10) $U_t(\top) - U_f(\bot) = 2\alpha - 1.$

According to (8)–(10),

(11) $U_t(h) = U_f(\bot) + 2\alpha - 1 + (1 - \alpha)cont(h),$

(12) $U_f(h) = U_f(\bot) - (1 - \alpha)(1 - cont(h)),$

and

(13) $U_t(h \vee \sim h) = U_f(\bot) + 2\alpha - 1.$[18]

The *expected* epistemic utility of accepting h is

$$(14) \quad \begin{aligned} E(h) &= p(h)U_t(h) + (1 - p(h))U_f(h) \\ &= p(h)(U_t(h) - U_f(h)) + U_f(h) \\ &= U_f(\bot) + \alpha p(h) - (1 - \alpha)(1 - cont(h)), \end{aligned}$$

where $p(h)$ is the (*a posteriori*) probability of h. The content measure *cont* can be thought of as being defined in terms of a probabilistic measure function m by

(cont) $cont(h) = 1 - (h),$[19]

but as (14) shows, m cannot be identical with the probability measure p; the content of a hypothesis is not necessarily inversely related to its *a posteriori* probability.[20]

V

According to the classical act-utilitarian view, an agent a ought to choose in a given situation an act A_i if and only if the consequences of A_i are better (more valuable) than those of any alternative action, and A_i is permissible for a if and only if it is at least as good as any alternative act:

(D.P) PA_i if and only if $V(A_i) \geq V(A_j)$ for every alternative A_j of A_i,

where $V(A_i)$ is the value of A_i. If we let $V(\sim A_i)$ be the maximum of the values associated with the alternatives of A_i, (D.P) can be expressed in the form

(D.P') PA_i if and only if $V(A_i) \geq V(\sim A_i)$.

In the present case the values of various doxastic attitudes can be defined as follows:

(15)   (a) $V(B_a h) = E(h)$
         (b) $V(B_a \sim h) = E(\sim h)$, and
         (c) $V(W_a h) = E(h \vee \sim h)$.

(15c) is motivated by the fact that if a withholds h, then (as far as h and its negation are concerned) he adds to his belief system nothing beyond the logical truth $h \vee \sim h$.

In (3) and (8)–(14), α and $1 - \alpha$ can be regarded as the relative weights assigned to truth and information (or content) as epistemic utilities: α is the weight assigned to the truth-factor, and $1 - \alpha$ the weight assigned to the information-factor of epistemic utility. If $\alpha = 1$, truth is the only relevant epistemic utility; in this case $U_t(h) = U_f(\perp) + 1$ and $U_f(h) = U_f(\perp)$. For example, if we let $U_f(\perp) = 0$, then $U_t(h) = 1$, $U_f(h) = 0$, and $E(h) = p(h)$. On the other hand, if $\alpha = 0$, only content matters: $U_x(h) > U_y(g)$ whenever $cont(h) > cont(g)$, regardless of the truth-value of h and g, and if $U_f(\perp) = 0$, $E(h) = -(1 - cont(h))$. (13) shows that if $\alpha \geq 1/2$, then $U_t(\top) \geq U_f(\perp)$, and consequently (according to (8) and (9)) no false hypothesis is epistemically preferred to any true hypothesis. In this case truth receives at least as much weight as content: content can never outweigh truth as epistemic utility. If $\alpha < 1/2$, some false hypotheses are epistemically preferred to some true hypotheses; for example, $U_f(h \ \& \ \sim h) > U_t(h \ \vee \ \sim h)$. Such a choice of α would obviously represent an unacceptable epistemic policy; thus we should require that $\alpha \geq 1/2$.

The assumption that $\alpha \geq 1/2$ implies

(16)   If $E(h) \geq E(h \ \vee \ \sim h)$,   then $E(h \ \vee \ \sim h) \geq E(\sim h)$.

According to (14),

(17) $E(h \vee \sim h) = U_f(\perp) + 2\alpha - 1$;

thus (according to (14) and (17)), the antecedent of (16) holds if and only if

(18) $\alpha p(h) - (1 - \alpha)(1 - cont(h)) \geq 2\alpha - 1$,

that is, if

(19) $\alpha p(h) + (1 - \alpha)cont(h) \geq \alpha$;

and the consequent of (16) is easily seen to be equivalent to

(20) $\alpha \geq 1 - (\alpha p(h) + (1 - \alpha)cont(h))$.

(19) implies (20) if and only if $\alpha \geq 1/2$; hence (16) is implied by the assumption that $\alpha \geq 1/2$.

(16) makes position P_3 impossible. In P_3, both $B_a h$ and $B_a \sim h$ are permissible, but $W_a h$ is not: this implies, according to (D.P) and (15), that the following three conditions are simultaneously satisfied:

(21)   (a) $E(h) \geq E(h \ \vee \ \sim h)$,
          (b) $E(\sim h) \geq E(h \ \vee \ \sim h)$,   and
          (c) $E(h) > E(h \vee \sim h)$ or $E(\sim h) > E(h \vee \sim h)$.

However, (16) and (21a) imply the negation of the second disjunct of (21c), and (16) and (21b) imply the negation of the first disjunct of (21c). Thus (16) implies principle (B1). This argument shows that principle (B1) follows from the assumption that content cannot outweigh truth as epistemic utility, that is, from the assumption that *no false hypothesis is epistemically preferred to any true hypothesis*.

A slightly stronger principle, viz. that *truth is always preferred to error*, corresponds to the assumption that $\alpha > 1/2$; in this case $U_t(h) > U_f(g)$

for any h and g.[21] If $\alpha > 1/2$, (19) implies

(22) $\alpha > 1 - (\alpha p(h) + (1 - \alpha)cont(h))$,

in other words, the assumption that $\alpha > 1/2$ implies

(23)   If $E(h) \geqq E(h \vee \sim h)$,   then $E(h \vee \sim h) > E(\sim h)$.

This means (according to (D.P) and (15)) that $B_a h$ and $B_a \sim h$ cannot both be permissible in the same situation. Thus (23) implies the following principle:

(B2) $PB_a h \supset \sim PB_a \sim h$.

(B2) is a stronger principle than (B1); it implies but is not implied by the latter. (B2) excludes both (P_3) and (P_1) from the set of possible normative positions. According to (P_1), all possible doxastic attitudes towards h are permissible for a, and the exclusion of this position may be intuitively justified as follows: Position P_3 is symmetric with respect to h and its negation. This symmetry may be due to the lack of relevant evidence concerning h and $\sim h$, or to the fact that the evidence for h is counterbalanced by that for $\sim h$. In both cases the proper doxastic attitude would seem to be withholding h, as indicated by position P_6.

VI

Roderick M. Chisholm has developed a system of epistemic logic which contains a binary relation of epistemic preferability (or reasonableness) between doxastic attitudes as a primitive epistemic concept.[22] This relation is characterized in terms of seven axioms; some of the axioms describe the general structural properties of the relation of epistemic preferability (such as asymmetry and transitivity), whereas others concern the relationships among different doxastic attitudes. The latter axioms include the following principle (Axiom 4; here the expression "more reasonable than" means epistemic preferability):[23]

(C4)   For every proposition h and subject S, if it is not more reasonable for S to withhold h than it is for him to believe h, then it is more reasonable for S to believe h than it is for him to disbelieve h.

Chisholm mentions the following instance of (C4): If agnosticism is not more reasonable than theism, then theism is more reasonable than atheism. If the relation of epistemic preferability is symbolized by 'R' and 'a' denotes an agent, this axiom can be expressed in the form

(C4') $\sim ((W_a h)R(B_a h)) \supset (B_a h)R(B_a \sim h)$.

If epistemic preferability is defined in terms of epistemic utilities in the obvious way, Chisholm's axiom is equivalent to the following principle:

(24)   If $E(h) \geqq E(h \vee \sim h)$, then $E(h) > E(\sim h)$,

which implies and is implied by (23). According to the analysis of epistemic preferability presented in Sections IV and V, Chisholm's fourth axiom is thus equivalent to the principle that truth is always epistemically preferable to error.

Some philosophers have argued that Chisholm's axiom (C4) is unacceptable, and attempted to find counter-examples to it. R. A. Imlay has argued that it is difficult to imagine circumstances in which a rational person could withhold the proposition "I am in excruciating pain".[24] According to Imlay, withholding this proposition would not be less unreasonable than believing it when it is in fact false. But then Chisholm's axiom would imply the patently unreasonable conclusion that even if a person is not in pain, it is more reasonable for him to believe that he is in excruciating pain than to disbelieve it; thus the axiom must be rejected. Imlay claims that tautological propositions generate similar counter-examples to (C4): for example, withholding belief in the law of identity is (according to Imlay) not epistemically preferable to disbelieving it, but believing it is nevertheless more reasonable than disbelieving it. In general, Chisholm's axiom fails to hold when withholding a proposition "possesses the same minimal degree of reasonableness as believing it".[25] More recently Melvin Ulm has presented essentially similar counterexamples to Chisholm's principle.[26] He has also argued that both believing that h and withholding h can have the same (minimal) degree of epistemic preferability if the negation of h is a self-evident truth.

We saw above that Chisholm's axiom can be regarded as equivalent to the assumption that truth is a more important factor of epistemic utility than content in the sense that truth (avoidance of error) is always epistemically preferable to error, regardless of the content of the propositions under consideration. It is clear that Imlay's alleged counterexamples to (C4) violate this condition: he assumes that there are cases in which $V(W_a h) = V(B_a h)$, although a has conclusive evidence for the negation of h. According to Imlay, "the assumption that it is always more reasonable to withhold a false proposition than to believe it" is false;[27] this is an explicit denial of the principle that truth (and avoidance of error) is always preferable to error, and Chisholm's reply to Imlay can be regarded as a convincing defense of the same principle. Chisholm notes that if one has conclusive evidence for a proposition, then disbelieving and withholding are both unreasonable, but "disbelieving is even more unreasonable than withholding, since in disbelieving, one *also* adds to one's stock of *false* beliefs.[28]

VII

Above we have seen that certain plausible assumptions about the objectives of rational belief expansion make positions P_1 and P_3 impossible. However, according to some recent theories of rational acceptance, P_2 and P_4 are also impossible doxastic positions. If the permissibility of a doxastic attitude is defined by (D.P), $B_a h$ and $W_a h$ have the same value in P_2, and in position P_4, $W_a h$ has the same value as $B_a \sim h$. In his book *Gambling With Truth* Isaac Levi notes that in practical decision problems "the options that tie for optimality are considered equally good and the agent is allowed free choice, or some other features of the decision situation are used to decide between them", but in "cognitive" decision problems free choice between the alternatives is illegitimate. If both $E(h)$ and $E(g)$ are maximal, the agent should, according to Levi, suspend judgment between h and g. This means that if $E(h) = E(h \vee \sim h)$, only $W_a h$ is a permissible doxastic attitude: Levi calls this rule "The Rule of Ties".[29] This rule involves the rejection of the purely "utilitarian" definition of doxastic permissibility. According to the Rule of Ties, $B_a h$ and $W_a h$ can never both be permissible in the same situation; thus it implies the validity of all instances of

$$(25) \qquad PA_i \supset \sim PA_j,$$

where $i \neq j$, and consequently the validity of all instances of

$$(26) \qquad PA_i \supset \sim P \sim A_i,$$

that is, the validity of

$$(B3) \qquad PA_i \supset OA_i.$$

Thus Levi's Rule of Ties leads to an extremely rigid ethics of belief: according to (26), a has never any "doxastic freedom": every knowledge situation involves exactly one acceptable doxastic attitude towards any proposition h. This rigidity does not appear totally implausible if we assume that the belief systems B may contain only propositions which are *certain* or *acceptable as evidence* (for other propositions). All propositions acceptable as evidence *ought* to be accepted as evidence (this is a version of Carnap's famous "Principle of Total Evidence"),[30] and if the evidential status of a proposition is open to reasonable doubt, then it should not be accepted as evidence, but ought to be withheld. However, (B3) does not hold for weaker senses of "belief": it is perfectly conceivable that a hypothesis is supported by strong but logically inconclusive evidence in such a way that neither believing it nor withholding it is unreasonable.

VIII

It is interesting to observe here that if $V(B_a h)$ is not measured solely in terms of the truth-value and the information-content of h, but depends also on various "practical" consequences of *a's believing that h* (or of the fact that a believes that h), (B1) and (B2) cannot be justified. Some philosophers who have discussed the ethics of belief have assumed that the value of a belief depends also on its non-cognitive effects on the person in question. For example, C. J. Ducasse has argued that a person has a right to believe anything which he thinks will be "a source of comfort, courage, and strength, and an inspiration to beneficience", provided that it does not conflict with his duty "to attend to evidence".[31] According to Ducasse, a belief cannot conflict with this duty if there is no "preponderance of evidence"; thus he assumes that in the absence of relevant evidence both $B_a h$ and $B_a \sim h$ (as well as $W_a h$) are permissible doxastic attitudes; this means that P_1 (and perhaps P_3 as well) is an acceptable normative position with respect to h. (It is not difficult to imagine situations in which both $B_a h$ and $B_a \sim h$ would have better effects on a person's life than doubt about the truth of h.) Thus (B1) and (B2) are acceptable principles of the ethics of belief only if the ethics of belief is regarded as a theory of cognitive or scientific rationality.*

University of Turku

* I am indebted to Professor Isaac Levi, Mr. Patrick Sibelius and Mr. Jari Talja for comments on the penultimate version of this paper.

NOTES

[1] Cf. H. H. Price, *Belief*, George Allen & Unwin, London 1969, Lecture 10: 'The Freedom of Assent in Descartes and Hume' (pp. 221-240); Descartes, *Meditations on First Philosophy*, in *The Philosophical Works of Descartes*, transl. by Elizabeth S. Haldane, and G. R. T. Ross, Cambridge University Press, Cambridge 1969, Vol. 1, pp. 174-178; and David Hume, *A Treatise on Human Nature*, ed. by L. A. Selby-Bigge, Clarendon Press, Oxford 1965, pp. 623-625.

[2] Peter Geach, *Reason and Argument*, Basil Blackwell, Oxford 1976, p. 3.

[3] A paper read in the Boston Colloquium for the Philosophy of Science, 1975, and presumably forthcoming in the Proceedings of the Colloquium. These types of belief change are also mentioned in Isaac Levi, 'On Indeterminate Probabilities', *Journal of Philosophy* 71 (1974), pp. 391-418 (see p. 396).

[4] Cf. Isaac Levi, 'Truth, Fallibility and the Growth of Knowledge', and 'Acceptance Revisited', in *Local Induction*, ed. by Radu J. Bogdan, D. Reidel, Dordrecht 1976, pp. 1-71 (cf. pp. 24-26).

[5] Here I am assuming that a cannot simultaneously believe h and $\sim h$, i.e. that $B_a h$ implies $\sim B_a \sim h$.

[6] Cf. Roderick M. Chisholm, *Person and Object*, George Allen & Unwin, London 1976, Ch. II.

[7] See Stig Kanger, 'Law and Logic', *Theoria* **38** (1972), pp. 105-134; Lennart Åqvist, 'Performatives and Veriafiability by the Use of Language', *Filosofiska Studier Utgivna av Filosofiska Föreningen och Filosofiska Institutionen vid Uppsala Universitet*, No. 14, 1972; Risto Hilpinen, 'On the Semantics of Personal Directives', in *Semantics and Communication*, ed. by Carl H. Heidrich, North-Holland, Amsterdam 1974, pp. 162-179; Ingmar Pörn, *Action Theory and Social Science*, D. Reidel, Dordrecht 1977, Ch. I.

[8] For the standard system of deontic logic, see Dagfinn Føllesdal and Risto Hilpinen, 'Deontic Logic: An Introduction', in *Deontic Logic: Introductory and Systematic Readings*, ed. by Risto Hilpinen, D. Reidel, Dordrecht 1971, pp. 1-35 (especially pp. 13-15).

[9] The present analysis of basic normative positions is adopted from Lars Lindahl, *Position and Change*, D. Reidel, Dordrecht 1977, Ch. 3 (pp. 36-92). Lindahl's analysis concerns the basic types of one-agent *legal* positions with respect to a given state of affairs (*op. cit.*, p. 92); here it is reinterpreted as a theory of the basic *doxastic* positions with respect to a proposition h.

[10] Cf. Ingmar Pörn, *Action Theory and Social Science*, pp. 4-8. The *Do*-operator used here corresponds to Pörn's D_a-operator.

[11] Lars Lindahl has given an example of this possibility in *Position and Change*, pp. 97-98.

[12] William James, *The Will To Believe*, Longmans Green and Co., New York 1897, p. 17.

[13] Alfred A. Knopf, New York 1967, pp. 75-108.

[14] Cf. *The Will to Believe*, p. 18. This Jamesian formulation of the cognitive objectives of belief expansion is slightly misleading: "Believe truth!" might be expressed more appropriately as "Believe informative hypotheses!"

[15] According to Levi, a "cognitive decision problem" (a problem of rational belief expansion) concerns the choice of the *strongest* accepted statement from a set of hypotheses which contains all "relevant potential answers" to a given problem or question. (See *Gambling With Truth*, Chs. II-IV.) From the standpoint of Levi's theory, the present discussion concerns a case in which the set of "relevant answers" is $\{h, \sim h, h \vee \sim h\}$. ($h \vee \sim h$ corresponds to the doxastic attitude of withholding h; cf. Section V below.)

[16] Here I am departing from Levi's use of α and $1 - \alpha$ in *Gambling With Truth*. In *Gambling With Truth*, p. 107, the utility of truth is weighted by $1 - \alpha$ and the utility of information by α, but in his paper 'Information and Inference' (*Synthese* **17** (1967), pp. 369–391) and in 'Epistemic Utility and the Evaluation of Experiments' (*Philosophy of Science* **44** (1977), pp. 368–386, see especially pp. 374–377), Levi has switched the roles of α and $1 - \alpha$ and denoted the weight of the truth-factor by α. I am following the latter usage.

[17] The common content of h and g is expressed by their disjunction: h and g may say something more than their disjunction, but both give at least the information carried by $h \vee g$. For a discussion of the *cont*-measure of information content and its relation to other measures of information, see Rudolf Carnap and Yehoshua Bar-Hillel, 'An Outline of a Theory of Semantic Information', reprinted in Y. Bar-Hillel, *Language and Information*, Addison-Wesley Publishing Company, Reading, Mass. and London, 1964, pp. 221–274 (esp. pp. 231–256).

[18] Epistemic utilities are assumed to be measurable on an interval scale; thus the zero point and the unit of the U-function can be chosen arbitrarily. In *Gambling With Truth* Levi takes $U_f(\perp)$ as the zero point of the U-function (p. 78). If $U_f(\perp) = 0$, then $E(h) = \alpha p(h) + (1 - \alpha)(1 - cont(h))$ and $E(h \vee \sim h) = 2\alpha - 1$. In 'Information and Inference' and in 'Epistemic Utility and the Evaluation of Experiments' Levi assumes that $U_f(\perp) = 1 - \alpha$; in this case $E(h) = \alpha p(h) + (1 - \alpha)cont(h)$ and $E(h \vee \sim h) = \alpha$. (Cf. 'Information and Inference, p. 379; 'Epistemic Utility and the Evaluation of Experiments', p. 377.)

[19] If $U_f(\perp)$ is chosen as the zero point of the U-function, $cont(h)$ is expressed in terms of m, and (14) is divided by α, then the expected utility of accepting h can be expressed in an especially simple form,

$$p(h) - qm(h),$$

where $q = (1 - \alpha)/\alpha$. Cf. 'Epistemic Utility and the Evaluation of Experiments', p. 377.

[20] In *Gambling With Truth* Levi defines *cont* in terms of a *uniform* measure function over the strongest (most informative) consistent hypotheses in which an investigator is interested in a given knowledge-situation. Levi calls such a set an "ultimate partition": the elements of an ultimate partition are mutually exclusive and jointly exhaustive hypotheses, and they imply all the other relevant hypotheses under consideration (cf. *Gambling With Truth*, p. 71, and note 15 above). According to this definition, we should assume here that $cont(h) = cont(\sim h) = 1/2$. Jaakko Hintikka and Juhani Pietarinen ('Semantic Information and Inductive Logic', in *Aspects of Inductive Logic*, ed. by Jaakko Hintikka and Patrick Suppes, North-Holland, Amsterdam 1966, pp. 96–112) and Risto Hilpinen (*Rules and Acceptance and Inductive Logic*, Acta Philosophica Fennica **22**, North-Holland Publishing Company, Amsterdam 1968, pp. 105–111) have defined *cont* in terms of the *a priori* probability of h; this definition was considered by Levi in 'Corroboration and Rules of Acceptance', *British Journal for the Philosophy of Science* **13** (1963), pp. 307–313, but was later rejected by him (cf. 'Information and Inference', p. 374; 'Acceptance Revisited', pp. 38–39).

[21] Various assumptions concerning α are discussed by Isaac Levi in *Gambling With Truth*, pp. 107–108, and in 'Information and Inference', pp. 378–379.

[22] Roderick M. Chisholm, 'The Principles of Epistemic Appraisal', In *Current Philosophical Issues: Essays in Honor of Curt John Ducasse*, ed. by F. C. Dommeyer, Charles C. Thomas, Springfield, Ill., 1966, pp. 87–104; *Theory of Knowledge* (Second edition), Prentice-Hall, Englewood Cliffs, 1977, pp. 12–15; Roderick M. Chisholm and Robert Keim, 'A System of Epistemic Logic', *Ratio* **15** (1973), pp. 99–115.

[23] This formulation is taken from *Theory of Knowledge* (2nd edition), p. 13.

[24] R. A. Imlay, 'Chisholm's Epistemic Logic', *Philosophy and Phenomenological Research* **30** (1969), pp. 290-293; see p. 291.

[25] Imlay, *op. cit.*, p. 292.

[26] 'Chisholm's Fourth Axiom', *Philosophical Studies* **27** (1975), pp. 57-61.

[27] Imlay, *op. cit.*, p. 290.

[28] Roderick M. Chisholm, 'On a Principle of Epistemic Preferability', *Philosophy and Phenomenological Research* **30** (1969), pp. 294-301; see p. 300. It is important to note here that the principle that truth is always preferable to error does *not* mean that it is never reasonable to accept false hypotheses. Even if h is false, it may be the case that $E(h) > E(h \vee \sim h)$. The principle means only that $U_t(h) > U_f(g)$ for any h and g (and consequently $U_t(h \vee \sim h) > U_f(g)$ for any g); this implies that it is never reasonable to accept a hypothesis if one has conclusive evidence for its negation. (Withholding such a hypothesis is always epistemically preferable to believing it; this is denied by Imlay and Ulm.)

[29] *Gambling With Truth*, pp. 84-85.

[30] Rudolf Carnap, *Logical Foundations of Probability*, The University of Chicago Press, Chicago 1951, pp. 211-212. Various interpretations of this requirement are discussed in Risto Hilpinen, 'On the Information Provided by Observations', in *Information and Inference*, ed. by Jaakko Hintikka and Patrick Suppes, D. Reidel, Dordrecht 1970, pp. 97-122.

[31] Quoted from Peter H. Hare and Edward H. Madden, *Causing, Perceiving and Believing: An Examination of the Philosophy of C. J. Ducasse*, D. Reidel, Dordrecht 1975, p. 150.

KAREL LAMBERT

EXPLANATION AND UNDERSTANDING: AN OPEN QUESTION?

I

My concern is the problem of the relation between scientific explanation and scientific understanding. What problem, you may ask. Didn't Hempel solve that one in 1965 in his book *Aspects of Scientific Explanation?*[1] Well, I'm not sure about that, as will be clear in a minute. But first I want to make some remarks on the importance of the problem.

A recurrent debate when I was doing graduate work in experimental psychology during the early 1950's was the following. Some tough-minded teachers and classmates declared the sole purpose of behavioral science to be the prediction and control of behavior; explanations, they said, ought to be left to the philosophers. So their preference for the alleged a-theoretical approach of B. F. Skinner was hardly surprising. With equal passion, their critics argued that a Skinnerian methodology reduces behavioral science to little more than sophisticated animal training, but that there is more to behavioral science than the prediction and shaping of behavior; the science of behavior seeks to provide understanding of the phenomena in its proper domain of interest, and understanding is the product of explanations, not of mere predictions. Quite naturally these scientists found theoretical approaches, such as cognitive theory, much more congenial. So it is a philosophical question of some scientific moment whether, and what kind of, understanding is provided by scientific explanation. In what follows Carl Hempel's conception of the matter is the initial topic of concern; it will serve as a foil for later speculative, and I'm afraid, rather impressionistic remarks. At any rate, appealing though Hempel's theory is, it seems to me, nevertheless, open to serious question. (I limit myself to the explanation and understanding of phenomena.)

Understanding a phenomenon is not, Hempel says, reduction to the familiar. It is rather a kind of objective insight. Witness, for example, the following passage:[2]

... the view that an adequate scientific explanation must, in a more or less precise sense,

29

R. Hilpinen (Ed.), Rationality in Science. 29–34.

effect a reduction to the familiar, does not stand up under close examination. . . . This view would seem to imply the idea that phenomena with which we are already familiar are not in need of, or perhaps incapable of, scientific explanation; whereas in fact science does seek to explain such "familiar" phenomena as the regular sequence day and night, and of the seasons, the phases of the moon, lightning and thunder, the color patterns of the rainbow, and of oil slicke, and the observation that coffee and milk or white and black sand, when stirred or shaken, will mix, but never unmix again. Scientific explanation is not aimed at creating a sense of at homeness or of familiarity with the phenomena of nature. That kind of feeling may well be evoked even by metaphorical accounts that have no explanatory value at all, such as the "natural affinity" construal of gravitation or the conception of biological processes as being directed by vital forces. Scientific explanation, especially theoretical explanation, is not this intuitive and highly subjective kind of understanding, but an objective kind of insight that is achieved . . . by exhibiting the phenomena as manifestations of common underlying structures and processes that confirm to specific, testable, basic principles.

Neither is scientific understanding merely psychological relief from perplexity, though no doubt it often has that result. According to Hempel, one has scientific understanding why a phenomenon occurred if one sees that, in virtue of certain laws, the phenomenon was to be expected, relative to certain circumstances.

The connection between scientific explanation and understanding why, thus, is quite direct. Deductive or inductive subsumption of a phenomenon under law affords the objective kind of insight that qualifies as understanding why. It does this by showing, deductively or inductively, that the phenomenon resulted from certain particular circumstances, in accordance with certain laws. Thus, in his book *Philosophy of Natural Science*, Hempel writes:[3]

Consider . . . the physical explanation of a rainbow. It shows that the phenomenon comes about as a result of the reflection and refraction of the white light of the sun in spherical droplets of water such as those that occur in a cloud. By reference to the relevant optical laws this account shows that the appearance of a rainbow is to be expected whenever a spray or mist of water droplets is illuminated by a strong white light behind the observer.

Three features of Hempel's conception should perhaps be emphasized. *First*, his account is intended to apply both to deductive explanations and inductive explanations. Thus he writes:[4]

. . . Any rationally acceptable answer to the question 'Why did event X occur?' must offer information which shows the X was to be expected – if not definitely, as in the case of *DN* explanation, then at least with reasonable probability.

Second, understanding why a phenomenon occurred is always relative to a certain set of circumstances; a phenomenon X may be expected to occur relative to one set of circumstances but not to another. Here is an example. Suppose a Geiger counter to be placed some considerable distance away from a milligram of polonium 218 in process of radioactive decay, and suppose the counter to click. Now relative to the decaying milligram of polonium 218 alone, the response of the Geiger counter is highly improbable (hence not to be expected) because of its distance from the polonium 218, but relative to a particle emitted from the mass of polonium 218 *striking the counter* the response of the counter is to be expected. *Third*, understanding why, in the Hempelian sense, does not imply that a prediction can be made. For example, consider a case attributed by Hempel to Michael Scriven.[5] The collapse of a bridge is explained as due to metal fatigue, and not to other factors such as excessive load. But knowledge of the cause of the collapse as due to metal fatigue, it is supposed, comes only after the collapse of the bridge. Then, under these circumstances the collapse of the bridge could not have been predicted from knowledge of the metal fatigue. Nevertheless it is still true to say that given prior knowledge of the metal fatigue as the sole cause of the bridge's collapse, the collapse of the bridge was to be expected. So understanding why does not reduce to prediction.

Objections to Hempel's theory of explanation have been abundant and varied for many years now. What has not been so clearly realized is that some of these very same objections affect Hempel's view about what understanding come to. Here is an example.

Suppose a certain alpha particle involved in the bombarding of a certain nucleus to tunnel through that nucleus. Understanding why this phenomenon occurred cannot be construed as an objective insight to the effect that the tunnelling through the nucleus of the alpha particle was to be expected given bombardment of the nucleus. For the fact of the matter is that such an event is highly unlikely, and indeed would not be expected under such circumstances. There is a probability of the order of 10^{-38} that such an event will ever transpire. (The example is due to Wesley Salmon.[6]) And if it is thought, as Salmon and others do, that non-Hempelian explanations of such unanticipated events are available, then it is clear that this sort of explanation does not provide the kind of objective insight that qualifies as Hempelian understanding.

The matter then can be put this way. Given Hempel's account of understanding why, a serious question may be raised whether explanations always provide such understanding. Here is a final example – due to Richard Jeffrey.[7] There are two cases. First, a coin is flipped once and a head occurs. Second, a coin is flipped 10 times and a head occurs. The explanation of a head occurring in the 10 flip case, according to Jeffrey, would cite the fact that a coin was flipped 10 times, that it was fair, that the procedure was a random process, and in addition cite certain general laws of probability.

A structurally similar and equally good explanation is forthcoming in the one flip case though the probability of getting a head in this case is 0.50 whereas in the 10 flip case it is "a bit over 0.999". Now Jeffrey thinks these probabilities have nothing essential to do with the explanation, but rather are merely good measures "of the strength of our expectation of" head occurring. If Jeffrey is right, then understanding of the Hempelian sort simply is not provided by many inductive explanations.

These questions about Hempel's theory of explanation and understanding evoke at least three responses. The first is that inductive statistical explanations ought to be junked. In fact, this option seems the one Hempel himself ought to adopt if one of his 1965 views is seriously taken into consideration. On page 367 of this book *Aspects of Scientific Explanation* Hempel notes that it is a "*condition of adequacy for any rational acceptable explanation of a particular event*" that it provide information showing that the event "was to be expected".[8] So provision of understanding, in the Hempelian sense of the word, appears to be a criterion of adequacy for any good scientific explanation, and since the inductive statistical candidates fail this test – at least if Jeffrey is to be believed – they should not be deemed legitimate scientific explanations.

Another response is to deny that all explanations of events provide understanding why – in particular, statistical accounts *à la* Jeffrey.[9] Inclining reasons for adoption of this option over the earlier one is the fact that Jeffreyan statistical answers seem to be quite proper responses to the question, How is phenomenon X to be explained? and at the same time, as Jeffrey himself at one place suggests, "strained" responses to the question, Why did phenomenon X occur?[10] Thus to the question How is the occurrence of one head on two tosses of a fair coin to be explained? it does not seem inappropriate to say "Well, you see, the process resulting in that outcome is a stochastic one yielding a probability of 1/2 on either toss, the coin was a fair coin, and the tosses were independent of each other; but there is no reason why a head occurred."

A third response compatible with the response just outlined is latent, I think, in Jeffrey's paper in the Salmon volume. It is that nondeductive explanations of the improbable do provide understanding, but of a different sort than is provided by deductive nomological explanations.

Deductive nomological explanations provide understanding, one may grant, in the sense of Hempel, or perhaps in the more familiar terms of providing an understanding of the causes of the phenomenon being explained. On the other hand, nondeductive explanations, Jeffrey suggests, provide understanding in the sense of providing an "understanding of the process" yielding the phenomenon.

Thus, Jeffrey writes:[11]

In the statistical case . . . I could rather speak of *understanding the process* which had the outcome, for the explanation is basically the same no matter what the outcome: it consists of a statement that the process was a stochastic one, following such and such a law.

So to the question, How did it come about that the ovum got fertilized by a sperm cell with Y genotype? the answer constituting the explanation yields the understanding that the usual process wherein spermatozoa unite with ova is a stochastic one in which the probability is $1/2$ that the fertilizing sperm cell will have the Y genotype. And this implies, says Jeffrey, the deeper insight that the process is not the usual causal process; in Aristotle's words that it is an "incidental" causal process rather than an actual causal process.[12]

One might put the point cryptically if vaguely in this way: explanations of the fertilization kind provide understanding of the sort reflected in answers to the explanatory request, How did phenomenon X come about? and not of the sort reflected in answers to the explanatory request, Why did phenomenon X occur?[13]

To sum up, I have tried here neither to analyze nor to answer a philosophical question, but rather to raise one. The deep problem concerns the exact relation between explanation and understanding; the surface problem concerns the kind of understanding, if any, provided by statistical explanations of improbable events, if such there be?[14] Three options were outlined, and which, if any, is the correct one I have no fixed opinion about. Nor do I have a clear grasp of just what is involved in each option. Still the problem seems both significant and important.

University of California, Irvine

REFERENCES

[1] Hempel, C. G.: 1965, *Aspects of Scientific Explanation*, The Free Press, New York.
[2] Hempel, C. G.: 1966, *Philosophy of Natural Science*, Prentice Hall, New Jersey, p. 83.
[3] Ibid:, p. 48.

[4] Op cit., *Aspects of Scientific Explanation*, pp. 367–368.

[5] Ibid., p. 371.

[6] Salmon, W. C.: 1971, *Statistical Explanation and Statistical Relevance*, University of Pittsburgh Press, Pittsburgh, p. 58.

[7] Ibid., pp. 23–24.

[8] Let me hasten to add that if push came to shove, what would most likely go in Hempel's theory would not be his doctrine of inductive statistical explanation; rather it would be the stated adequacy condition. But even if Hempel cannot be listed among those who ultimately reject the view that there are statistical explanations, proponents are not hard to find. In fact, they are geographically very close; for example, my colleague Kit Fine, who resides only four doors away from my office, is one such person. (Since this was written, Kit Fine, regrettably, has become my former colleague, though there is no causal connection.)

[9] Informally, Terence Parsons has proposed such a view to me. Whether he would keep the same belief after sustained reflection I cannot say. I add Parsons' informal observation only in the interest of undoing a possible strawman charge – something that can be accomplished by a single example.

[10] Op. cit., *Statistical Explanation and Statistical Relevance*, p. 24.

[11] Ibid., p. 24.

[12] It is necessary, no doubt, that the causes of what comes to pass by change be indefinite; and that is why chance is supposed to belong to the class of the indefinite and to be inscrutable to man, and why it might be thought that, in a way, nothing occurs by chance. For all these statements are correct, because they are well-grounded. Things *do*, in a way, occur by chance, for they occur incidentally and chance is an *incidental cause* [see 195a 28–195b 3]. But strictly it is not the *cause* – without qualification – of anything; for instance, a housebuilder is the cause of a house; incidentally, a flute-player may be so. [*Physics* II.5.197a 8–14]. (I want to thank John Vickers for digging up this pair of passages from Aristotle. Further, I appreciate the comments he expressed on an earlier draft of this address.)

[13] It might be objected that the two questions are not really distinct; for example, both, it might be said, "presuppose" causal answers. Though I am aware of the vague and perhaps even ultimately unhelpful character of these two questions, the particular charge in question is, I think, unjustified. Whereas I am willing to grant that "Why did X occur?" does presuppose a causal answer – or at least a close facsimile, I don't think the same is true of "How did X come about?" (or "How did X result?"). It makes sense to respond to this latter question that a certain event came about as the result of, or simply resulted from, a stochastic process, and these processes simply are not causal processes.

[14] Throughout I have supposed that *what* is being understood, if anything at all, is the same in the case both of deductive and inductive explanations, namely, the event, happening, or what have you soliciting the explanation. But that assumption may indeed be unwarranted. It may very well be that reflection will sustain the intuition of some who have discussed the matter with me that only in the case of deductive explanations can one legitimately say that what is understood is the explained occurrence.

ADRIENNE LEHRER

THE EMPIRICAL INVESTIGATION OF SYNONYMY
AND THE IMPLICATION FOR SCIENCE*

Philosophers, linguists, and psychologists have long been interested in synonymous expressions, often for rather different reasons. The philosopher's interest in synonymy is related to his interest in logical truth; if two expressions are synonymous, then a sentence in which one expression is exchanged for a synonymous expression will retain the same truth value. Traditional philosophical problems like analyticity, contradiction, etc. are tied up with this problem. Linguists, especially lexicologists and lexicographers, have been interested in synonymy. The task of the lexicologist is to show what the relationship is between expressions in a language, and synonymy is one of these relationships. Psychologists have been interested in similarity of meaning rather than synonymy *per se* in order to determine how people learn, store, and use information, including information about speakers' knowledge of their language. Notions like generalization and concept formation often refer to perceived similarity, and similarity of meaning is thought to reflect perceived similarity of stimuli.

The theoretical treatment of 'synonymy' as well as the methodology used for determining synonyms varies from writer to writer as well as from discipline to discipline. Some philosophers, e.g. Katz (1972), are more interested in how to represent synonyms than in devising tests for whether there are any synonyms in a given language. Only a few classical examples, such as *bachelor* and *unmarried man*, are formalized.

Studies in psychology have been done for various purposes, and some of these studies were done or at least begun under the behavioristic influence when reflection, introspection, or self-conscious reports of meantalist concepts like meaning were suspect. Osgood et al., using the Semantic Differential, acquired information on affective meaning (subjective connotation) rather than referential or cognitive meaning. Deese, using word association tests, found that a stimulus word usually produces an antonym, and often a synonym or a superordinate or subordinate word, but he did not distinguish the kinds of responses. More recently, studies in which subjects classify stimulus words into clusters, or decide that word A is more similar to B than to C have been used,

35

R. Hilpinen (Ed.), Rationality in Science. 35–50.
Copyright © 1980 by D. Reidel Publishing Company.

employing sophisticated statistical analyses; but so far most of the stimuli used have not been synonyms (Miller, Fillenbaum, and Rips).

It has been argued that there are no true synonyms, where 'synonymy' is defined as substitutibility preserving meaning. (See Goodman.) Also linguists and rhetoriticians have argued that nearly synonymous expressions may differ in subtle nuances or that the meaning of a word may vary according to context. While this may be true, it is useful to adopt a looser definition of synonymy – something short of complete substitutability – such as great overlap in many contexts or great similarity of meaning.

Quine (1953), in part for epistemological reasons, has been one to challenge theories of language that hold to an analytic-synthetic distinction and which use synonymy as a basic (or derived) concept in the theory. In place of synonymy, Quine (1953a) proposes a sliding scale of similarity from very similar to less similar. He writes, "Quite possibly the ultimately fruitful notion of synonymy will be one of degree: not the dyadic relation of *a* is synonymous with *b*, but the tetradic relation of *a* is more synonymous with *b* than *c* with *d*" (Quine, 1953a). This study is an attempt to test Quine's hypothesis.

Earlier work that I had done on speakers' ability to distinguish homonymy from polysemy suggested an experimental test that could be adapted to studying synonymy (Lehrer, 1974). Subjects were given two items and asked to judge their similarity on a 5 point scale, where 1 was to be used for very similar and 5 was to be used for unrelated items. There were two sets of items. One set consisted of phonologically identical items with glosses, such as:

orange – a color	*bag* – paper sack	*check* – bill for meals
orange – a fruit	*bag* – a purse	*check* – pattern of squares

The other set of items consisted of phonologically different pairs:

apple	*purse*	*bill*
red	*sack*	*plaid*

Subjects were given the test and were then retested on the same material one month later. Results showed that speakers' judgements were highly uniform for the pairs of items that were judged most similar or most different (as measured by standard deviation). Moreover, judgements on the retest were highly consistent both for individual speakers and for the groups on both the most similar pairs of items and on the unrelated items, whereas judgements were less consistent on the other pairs of items. These results would suggest that if synonyms are defined as pairs (or classes) of expressions that are very similar in meaning, judgements about their similarity should be relatively uniform from speaker to speaker and the judgements of speakers should be consistent over time, whereas items that partially overlap in meaning will vary more from speaker to speaker and speakers will not be consistent in their judgements. However, the number of items used in the previous experiment was too small to draw this conclusion.[1]

There still remain problems in devising a suitable methodology for determining similarity of meaning. Scaling was used in the homonymy-polysemy experiment, since it is considered a standard technique for such purposes (Miller, 1971), and subjects seem to understand what is required of them and perform the task easily and willingly. However, this method lacks subtlety, and it is not always clear to subjects exactly what the experimenter is looking for; nor is it always clear from studying the results what the subjects are thinking about and responding to. On the other hand, the use of intuitions of the researcher and his or her colleagues lacks reliability and does not account for the considerable individual variation among speakers.

In philosophy, Naess (1953) has investigated synonymy empirically and has discussed in great detail methods for determining synonymy in texts or preparing questionnaires for subjects. He has concentrated on different ways of formulating questions, but in most cases subjects must respond with either a *yes* or *no* answer. My studies differ from those of Naess in that scaling responses are elicited rather than *yes* or *no*, and the stimuli selected include items that are not complete synonyms but which share some overlap of meaning. Moreover, what is measured is consistency and variability of responses rather than the responses *per se*.

Since many writers, philosophers, linguists, etc. have warned that there are few 'real' synonyms, an attempt was made to find synonyms empirically – that is, by using the scaling technique, such as considering as synonyms those pairs of words that were judged as 1 or 2 on a five point scale.[2] Tentative candidates were selected on the basis of dictionaries of synonyms, Roget's Thesaurus, and the intuitions of the experimenter.

A pilot study was done to determine three things: (1) whether to use a

5 or 7 point scale for judging similarity; (2) whether to use words in isolation or in sentence contexts; and (3) how subjects would judge antonyms and converse terms as compared to unrelated terms.

A seven point scale had been used in the pilot test, but results showed that this provided too many choices and that a 5-point scale would be preferable, as in the homonymy-polysemy experiment.

In determining the effect of sentence context on judgements of similarity, subjects were given pairs of words to rate on a 7-point scale, such as *big-large*, *boat-ship*, or *warm-hot*. At a different session, they were given pairs of sentences containing those words, for example

> Sally lives in a big house
> Sally lives in a large house.

In addition, some sentence contexts were selected to maximize the differences, such as

> Sally is my big sister
> Sally is my large sister.

In the latter pair of sentences, subject judged the sentences to have quite different meanings. This is due to the fact that *big* has a second meaning – 'older' in addition to 'large size'. However, except for those contexts involving the polysemy of words, responses to pairs of words in isolation did not differ significantly from those in neutral sentences contexts. Those pairs of words that did differ were not used in later experiments. (The problem of polysemy is discussed below.) A few sentence contexts produced responses where judgements were closer on sentences than words in isolation, such as

> He added up the numbers correctly.
> He added up the numbers accurately.

Lyons (1963) has proposed that antonyms are closer in meaning than unrelated words, since they share some features and belong to the same semantic field, e.g. *brother* and *sister* are kin terms, *tall* and *short* are height terms. It was hypothesized that subjects would judge such pairs as closer in meaning than totally unrelated items, such as *love-exaggerate* or *painter-grapefruit*. However, results showed that the mean scores on the antonyms were not significantly lower than on the unrelated terms. This does not necessarily mean that subjects did not see any similarity but rather that two kinds of difference were involved –

contrariness and unrelatedness. Subjects were required to collapse these two dimensions into a one-dimensional scale. Therefore, in the later testing, antonyms and converses were not used.

Four questionnaires were presented to students in an introductory linguistics class at the University of Arizona. The interval between questionnaires was three to four weeks. The first and third questionnaires asked subjects to judge the similarity of two words on a 5-point scale. The second and fourth questionnaires asked subjects to decide which of two pairs of words were closer in meaning, i.e. whether A and B were more similar or C and D.

In the first test, 45 pairs of expressions were presented to subjects in a randomized order. The expressions included probable synonyms (e.g. *big-large*), a word and a synonymous phrase (*kill-cause to die*), words with partial overlap (*cow-animal*), and unrelated terms (*green-sad*).[3] There was an attempt to have at least 7 or 8 pairs in each category.[4] The same items were presented to subjects about six weeks later, but the items were in a different order.

The instructions presented to the subjects were as follows:

> Below are pairs of expressions (usually words, sometimes a word or phrase). You are to judge how close in meaning each pair is. If the items are synonyms, place an X over the 1. If the items are completely unrelated, place an X over the 5. Use 2, 3, 4 for intermediate judgements; e.g. 2 for items that are close in meaning, but are not quite synonymous, 3 for less similarity, etc. You should quickly skim over the list before making your judgements to get some idea of the range of items.

The results are presented in Table I. The first column presents the average of the means on both the test and the retest, and the items are

TABLE I

	Mean	$\hat{s}$	t	Absol. Change	% of S ($N = 30$)
pail – bucket	1.253	0.46		8	23
puppy – young dog	1.255	0.454	*	6	21
aid – help	1.265	0.518		16	40
kitten – young cat	1.288	0.51		6	24
small – little	1.302	0.485		9	30
big – large	1.319	0.96		9	32
rob – steal	1.387	0.602		9	30
fast – swift	1.436	0.570		9	31
tasty – having a good flavor	1.495	0.663		15	48
accurately – correctly	1.625	0.718		14	37
grief – sorrow	1.660	0.729		21	60
thin – skinny	1.713	0.655		12	37
terrible – horrible	1.744	0.914		16	33
deep fry – French fry	1.813	1.017		27	66
brook – stream	1.815	0.742		12	33
distruct – be suspicious of	1.843	0.831	**	25	66
king – emperor	1.867	0.903		19	43
hunt – search	1.888	1.923		21	47
violet – purple	1.946	0.737		20	60
boat – ship	2.065	0.637	*	16	37
level – flat	2.16	0.803	**	25	60
rough – rugged	2.208	0.994		19	53
melt – dissolve	2.272	1.119		17	50
bake – cook	2.331	0.714		12	37
kill – cause to die	2.413	1.29		23	59
tulip – flower	2.416	1.451		19	55
river – creek	2.496	0.589		19	50
invent – discover	2.564	0.435		21	47
hand – paw	2.609	1.032		26	60
cow – animal	2.766	1.007		28	70
hot – warm	2.848	1.21		21	50
pink – red	2.869	1.109		24	67
soft – smooth	2.972	1.115		21	50
slowly – casually	3.121	0.95		17	47
wisdom – information	3.554	0.952		19	40
pungent – savory	3.742	1.181		21	45
read – be aware of	3.989	0.901		19	30
warmth – innocence	4.189	0.97		23	60
performance – register	4.627	0.778		12	30
know – doubt	4.717	0.575		7	20
green – sad	4.826	0.409		5	13

	Mean	$\hat{s}$	t	Absol. Change	% of S ($N = 30$)
hold – drive	4.836	0.41		5	17
sit – cry	4.927	0.234	*	7	17
chair – husband	4.938	0.283		3	7
tall – happy	4.938	0.461		3	7

Column 4 – Difference between Means on Test and Retest.

  * = $p < 0.05$

** = $p < 0.01$

arranged in order of similarity.[5] A Pearson Correlation between the means on both tests was 0.989, showing that the means were fairly stable from test to retest. The third column in Table I shows those items for which the means on the two tests were significantly different, using t-test. One asterisk shows a difference at the 0.05 level of significance, and two asterisks show a difference of 0.01. The only major break in the ranking occurs between *warmth-innocence* with a mean of 4.189 and *performance-register* with a mean of 4.627, a difference of almost half a point. This break could operationally establish the difference between unrelated pairs and related pairs. Therefore, we can operationally define unrelated pairs of words as having a mean of 4.5 or greater. There are 7 such pairs.

Turning to the most similar items, there is no sharp break in the ranking, but perhaps we can operationally define synonyms as those items that have a mean of 1.5 or less. According to this criterion, there would be 9 such pairs.

It was hypothesized that unrelated pairs and synonyms would have a smaller standard deviation than the other items. It was further hypothesized that responses to these items would be more uniform and consistent. Column 3 gives the average standard deviation for each item on the two tests. Of the items with $\hat{s}$ of 0.5 or less, 5 are unrelated. The probability of this result (binominal distribution) is < 0.004. Of these items, 3 are synonyms ($p < 0.15$). Combining the classes, of the 9 items with $\hat{s}$ of less than 0.5, 8 are unrelated or are synonyms ($p < 0.007$).

An alternative method of computing significance is as follows: If we look at the 10 items with the smallest $\hat{s}$, 5 are unrelated ($p < 0.006$) and 4 are synonyms ($p < 0.067$). Combining synonyms and unrelated items:

Of the 10 items with the smallest $\hat{s}$, 9 are unrelated or synonyms ($p < 0.0002$).

To measure individual variation from one test to the next, the questionnaires from 30 subjects were randomly selected. For each item, the difference was calculated. For example, if a subject rated *big-large* as 1 on the first test and 3 on the retest, a value of 2 was recorded. If a second subject gave a 2 on a first test and a 1 on the retest, a value of 1 was recorded, giving a total of 3. Column 5 gives the cumulative totals of change (disregarding the direction of change) for the 30 subjects. When these values are ranked from lowest to highest, it can be seen that the items with the lowest values are unrelated. Of the 7 items with the lowest value, 6 are unrelated ($p < 0.001$). Of the 13 items with the lowest value, 7 are synonyms ($p < 0.005$). Column 6 gives the percentage of subjects who changed their judgement on an item between the test and retest. Of the 7 scores with the smallest number of subjects changing their judgement, 6 are unrelated ($p < 0.001$). Of the 15 items with the smallest change, 7 are synonyms ($p < 0.015$). Therefore, the results show that the judgements are most consistent on the unrelated pairs of expressions, and the second most consistent on pairs of expressions are those closest in meaning – the synonyms.

JUDGING PAIRS OF PAIRS

The second task consisted of presenting subjects with 22 pairs of pairs in a randomized order. Subjects were to decide which of the pairs was closer in meaning (Quine's test). The hypothesis was that if one pair of words was close in meaning and the other pair not close in meaning there would be high intersubject agreement on which pair to select. However, if both pairs were synonyms, then there would not be significant agreement on which pair was closer in meaning. Secondly, it was hypothesized that when the items were presented to subjects at a later date, individuals would make the same judgements on pairs in which there were significant differences between the similarity of pairs, but they would change their judgement more frequently when both pairs were synonyms. The task was repeated about six weeks later with 20 of the same items in a different order.

The instructions presented to the subjects were as follows:

Below are two pairs of words (or expressions). You are to

decide which pair is closer in meaning. For example, if given the two pairs

small little   green happy

you feel that the meanings of *small* and *little* are closer to each other than the meanings of green and happy, draw a circle around the pair *small little*.

The means used in the first task were used in the selection of pairs of pairs. Some items were selected that were maximally different such as a pair of synonyms and a pair of unrelated terms. Other pairs were matched to be maximally similar.

RESULTS

The results of the pairs of pairs test are given in Table II. The items are arranged in order of subject agreement. The results of the first test are presented in the left-hand side of Table II. The first column gives one member of each pair, followed by the mean for that item as determined by the scaling task. The third column shows the number of subjects that circled that pair. The fourth column presents the second pair, matched with the pair of column 1, followed by the mean (determined by the scaling task). Column 6 gives the number of subjects who circled the pair in column 4. The total number of subjects responding to the item is presented in column 7 (maximum = 52). Column 8 provides the probability (binominal distribution) of such results by chance.

Two things can be noted. The means for the pairs derived from the scaling task are a good predictor of which pair of pairs will be selected as most similar, reinforcing the validity of scaling as a tool. The only case in which the prediction was wrong was for *distrust – be suspicious of*, with a mean of 1.875 and *tasty – having a good flavor*, with a mean of 1.496. *Distrust – be suspicious of* was selected as closer in meaning more often, although this result was not statistically significant. The second point to note is that when the pairs of pairs were very close, as with *puppy – young dog* and *kitten – young cat*, a number of subjects refused to make a choice, and many who did wrote on the questionnaire that the choice was arbitrary.

In the retest, the ninth column provides the number of subjects who circled the pair in column 1, while the tenth column presents the

TABLE II
Pair of pairs

First Test								Retest					
1	2	3	4	5	6	7	8	9	10	11	12	13	14
	Mean	No.		Mean	No.	Total S's	p	No.	No.	Total	p	No. of Changes	% of Change N = 30
level - flat	2.16	52	happy - tall	4.938	0	52	> 0.001	44	0	44	> 0.001	0	0
slowly - casually	3.121	50	know - doubt	4.717	2	52	0.001	42	2	44	0.001	1	3
grief - sorrow	1.66	49	wisdom - innocence	4.189	2	51	0.001	43	1	44	0.001	2	7
kill - cause to die	2.413	48	read - be aware of	3.989	4	52	0.001	41	3	44	0.001	2	7
violet - purple	1.946	46	pink - red	2.864	6	52	0.001	38	5	43	0.001	2	7
wisdom - information	3.554	39	performance - register	4.627	8	47	0.001	38	6	44	0.001	5	17
pail - bucket	1.252	41	brook - stream	1.815	11	52	0.001	25	18	43	ns	8	27
boat - ship	2.065	40	hand - paw	2.609	12	52	0.001	38	6	44	0.001	6	20
deep fry - French fry	1.813	40	melt - dissolve	2.272	12	52	0.001	32	11	43	0.01	9	30
rough - rugged	2.208	38	hot - warm	2.848	14	52	0.001	34	10	44	0.001	8	27
king - emperor	1.813	38	river - creek	2.542	14	52	0.001	40	4	44	0.001	5	17
hunt - search	1.888	37	melt - dissolve	2.272	14	51	0.005	29	15	44	0.05	10	33
hold - drive	4.836	33	sit - cry	4.854	14	47	0.01	30	13	43	0.01	7	23
big - large	1.319	36	terrible - horrible	1.744	16	52	0.01	30	14	44	0.05	12	40
tulip - flower	2.416	33	cow - animal	2.766	16	51	0.01						
small - little	1.302	33	thin - skinny	1.713	19	52	0.05						
distrust - be suspicious of	1.875	31	tasty - having a good flavor	1.496	21	52	ns	26	18	44	ns	11	37
soft - smooth	2.972	30	pungent - discover	3.742	21	51	ns	22	21	43	ns	6	20
bake - cook	2.331	28	invent - discover	2.564	23	51	ns	27	17	44	ns	9	30
puppy - young dog	1.255	24	kitten - young cat	1.288	20	44	ns	30	13	43	0.01	14	47
fast - swift	1.436	28	accurately - correctly	1.625	24	52	ns	20	24	44	ns	4	13
aid - help	1.265	26	rob - steal	1.387	24	50	ns	23	20	43	ns	12	40

TABLE III

Relationship between results of the two tasks

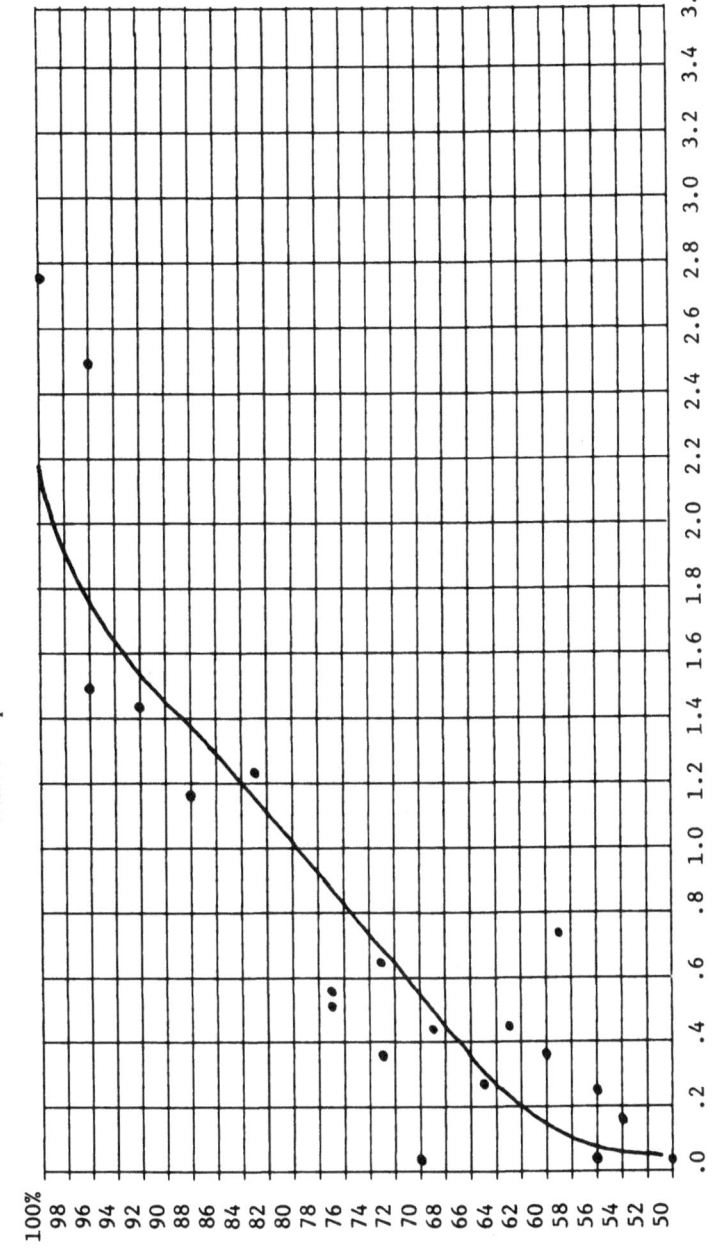

Difference Between the Means

Percentage of Consistent Responses

number who circled the pair in column 4. The total number of subjects responding to that item is given in column 11 (maximum = 44). The two questionnaires (test and retest) were compared for 30 randomly selected subjects to determine the changes in choice made on the retest. The 13th column gives the number of changes made, and the last column gives the percentage of subjects changing the judgements on the retest. It can be seen that there were the fewest number of changes when the pairs of items differed significantly in their similarity; the number of changes increased as the difference between the means decreased. The Pearson Correlation between the number of changes (column 13) and the responses in column 9 was 0.64.[6]

Table III provides a summary of the relationships between the results on the two tests. The vertical axis presents the percentage of subjects who circled the same pair of words on the test and retest of task 2. The horizontal axis presents the difference between the means of the two pairs from task 1. As can be clearly seen, as the difference between the means increases, the consistency of the responses increases.

DISCUSSION

What is the relevance of these tests to Quine's proposals? In some ways they show that Quine is right in arguing that there are degrees of similarity. Subjects were, after all, able to scale pairs of words and make judgements on which of two pairs was more similar. However, there are limits to scalability. The pairs of items seem to fall in clumps such that judgements of similarity within those clumps is not reliable. Responses are most uniform and consistent to those pairs that are unrelated in meaning. And the next most stable group, based on uniform and consistent responses, is to those pairs that are most similar in meaning – the pairs that have normally been considered synonyms.

We have seen that defining synonymy in terms of complete substitutability is too strong. A weaker definition would be to define synonymy as substitutability in a wide variety of contexts – including linguistic contexts and the application of words to things, properties, states, etc.

A complication which must be mentioned here is the polysemous nature of words, a problem which Quine does not mention. When two words that are often interchangeable are placed in a context where meaning changes, e.g. *my big sister* versus *my large sister*, the change may be due to the fact that one (or both words) have multiple meanings, and different meanings are involved in different contexts – an example of equivocation, if you like. Therefore, we might wish to define synonymy as relative to senses of words. Of course, we are still faced with the traditional lexicological problem of determining senses of words.

While the problems of determining synonmy are well known, I think that Quine underestimates the difficulty of analyzing the notion of similarity. Let us consider some pairs of words that were judged to be somewhere between synonymous and unrelated on the scaling test. *Pink* and *red*, for example, were judged as similar, and they are similar in that they refer (among other things) to parts of the color chart which are close to each other. But the words are not likely to be substitutable for each other in a wide variety of contexts. Or consider *hand* and *paw*. These two refer to loosely analogous parts of different kinds of animals, but their interchangeability is somewhat limited. Therefore, what these tests show, I believe, is that there is something special about the pairs of words that are most similar in meaning. These we can call synonyms, weakening the definition often given by philosophers. And I think that we can predict more about the behavior of synonyms than of other similar nonsynonymous words. The notion of similarity is not more useful or clearer than that of synonymy.

IMPLICATIONS OF THE RESULTS

In philosophy, Quine has challenged the validity of a distinction between analytic and synthetic statements. Synonymy yields analyticity, so that if there are synonyms, as I hope to have shown, then there are analytic sentences. And if there are analytic sentences, then there are important implications for science. If my claims are correct, the manner in which one could determine the truth of the following two sentences is different:

1.   Bachelors are unhappy.
2.   Bachelors are unmarried.

To discover the truth of 1, the investigator would have to do a psychological profile of a sample of bachelors, but to investigate the truth of 2, he would not.

In linguistics, since synonymy is one of the major semantic relationships that lexicologists investigate, it is important to know that this theoretical term has denotation and is not scientifically useless, like 'ether'.

The final reason is less obvious than the two given, but it is connected to more practical concerns. In many of the social sciences the main source of data is what people say about themselves and about others. For example, some aspects of the psychology of perception, social psychology, and psychiatry drawn on introspective reports. Since not all patients or subjects will report their feelings and experiences in exactly the same words, it is important to know what expressions are equivalent. Suppose patient A reports to the psychiatrist,

> 'My mother was economical but my father was miserly.'

while patient B reports

> 'My mother was thrifty but my father was stingy.'

If the psychiatrist can reasonably infer that both patients are conveying the same information, there may be significant generalizations that can be made, provided of course that *economical* and *thrifty* are synonymous as well as *stingy* and *miserly*.

An area of research in which I have been working most recently is in the domain of wine terminology, in which speakers encode into language the perceptual properties of wine – especially taste, smell, and feel. The synonymy in this field is partial and restricted, but to make appropriate generalizations, and I think it is legitimate to do so, one has to infer that if A says 'This wine is weak', while B says 'This wine is thin', they are talking about the same property. It is neither practical nor necessary to go through every predicate and to get every subject or patient to use a standardized vocabulary.

In short, *synonymy* is a useful and behaviorally valid concept, and contrary to Quine's criticisms, there is a good reason not to give it up. What Quine proposes in its place – similarity of meaning – is vaguer and less useful.

University of Arizona, Tucson

NOTES

* I wish to thank Keith Lehrer, Christine Tanz, and Robert M. Harnish, and participants at the conference for their helpful and insightful discussions of both the conceptual and experimental issues involved in synonymy. All shortcomings are mine.

[1] A second result of the experiment was that in comparing phonologically identical pairs (e.g. the two senses of *orange*) with matched phonologically different pairs (*apple* and *red*) the phonologically identical items were judged as closer in meaning than the other.

[2] Suppes (1973) has suggested that synonymy can be considered in terms of the geometrical concept of congruence. Since congruence can be strong or weak, depending on what must remain invariant (shape, size, orientation, etc.) considering sameness (or closeness) of meaning could be dealt with in terms of degrees or kinds of congruence.

[3] All pairs of words selected belonged to the same part of speech, since Angelin (1970) has shown that adults will judge words belonging to different parts of speech as less similar than those belonging to the same part of speech, even when semantic content is comparable.

[4] Tversky (1977) shows that similarity judgements will be influenced by the selection of stimuli. In the present experiment, for instance, if most of the stimuli were pairs of words that were very similar in meaning, it is likely that subjects would use some higher numbers for near synonymys, whereas if most of the pairs were not close in meaning, the few near synonyms would be more likely to be judged as a 1.

[5] A questionnaire was discarded if a subject did not use the full range of numbers available. In test 1, one questionnaire was discarded because the subject used no 4's or 5's, and another was discarded because he used no 1's. In test 3, one questionnaire was discarded for having no 1's.

The response to an item on a questionnaire was discarded if it was two or more numbers away from all the rest of the responses. For instance, a single or double judgement of 5 was discarded when all the other judgements were 3 or less. Eight scores were discarded in test 1, or 0.003; 4 were discarded in test 3, or 0.002. When the subjects' judgements were compared on the two tests, it seemed that the discarded responses were indeed mistakes.

[6] The values of column 9 were converted to percentages (column 9 ÷ column 11), since not every subject responded to every pair.

REFERENCES

Angelin, J. M.: 1970, *The Growth of Word Meaning*, MIT Press, Cambridge, Massachusetts.

Deese, J.: 1965, *The Structure of Associations in Language and Thought*, Johns Hopkins University Press, Baltimore, Maryland.

Fillenbaum, S. and A. Rapoport.: 1971, *Structures in the Subjective Lexicon*, Academic Press, New York, New York.

Goodman, N.: 1949, 'On likeness of meaning', *Analysis* 10, 1–7.

Katz, J.: 1972, *Semantic Theory*, Harper and Row, New York, New York.

Lehrer, A.: 1974, 'Homonymy and polysemy: measuring similarity of meaning', *Language Sciences* 3, 33–39.

Lyons, J.: 1963, *Structural Semantics*, Blackwells, Oxford, England.

Miller, G.: 1971, "Empirical methods in the study of semantics', in *Semantics: An Interdisciplinary Reader*, ed., D. Steinberg and L. Jacobovits, 569–585.

Naess, A.: 1953, *Interpretation and Preciseness*, I Kommisjon Hos Jacob Dybwad, Oslo, Norway.

Osgood, C. et al.: 1953, *The Measurement of Meaning*, University of Illinois Press, Urbana, Illinois.

Quine, W. V. O.: 1953a, 'The problem of meaning in linguistics', in *From a Logical Point of View*, Harvard, Cambridge, Massachusetts, 47–64.

Quine, W. V. O.: 1953b, 'Two dogmas of empiricism', in *From a Logical Point of View*, 20–46.

Rips, L. et al.: 1973, 'Semantic distance and the verification of semantic relations', *Journal of Verbal Learning and Verbal Behavior*, **12**, 1–20.

Suppes, P.: 1973, 'Congruence of meaning', Presidential address delivered at 47th Annual Meeting of the Pacific Division of the American Philosophical Association.

Tversky, A.: 1977, 'Features of similarity', *Psychological Review* **84**, 327–352.

KEITH LEHRER

A MODEL OF RATIONAL CONSENSUS IN SCIENCE

In a series of papers, I articulated a model of rational consensus concerning questions of science and social policy.[1] The problems we are discussing at this conference concern rationality in science and the application of scientific results. They fall within the domain of the model. In this paper, I shall sketch a simple model, consider some objections, and propose a modification of the method recently articulated by my collaborator, Carl Wagner. I shall begin with a normative application of the model and then speculate about whether some empirical phenomena might conform to the model as well.

THE SIMPLE MODEL

Imagine that people disagree about some scientific question, for example, one concerning the shape of the sun in our solar system. In this case, different scientists might assign different probabilities to the hypothesis that the sun is oblate. Let us call this the hypothesis 'B' and the chief competitor, the hypothesis that the sun is round, the hypothesis 'R'. Now even scientists who more or less agree about the scientific data and the merits of theoretical extrapolation from the data might assign a variety of probabilities to B and R. How might we proceed to find some probability assignment that would constitute a reasonable summary of all the relevant information?

The first step would be to insure that all of the probability assignments we consider are consistent with the experimental data. Thus we must impose a *consistency condition* on the probability assignments of members of the group to the effect that the probability assignments of all members of the group are consistent with all *objective* statistical and frequency data obtained from precise observation and controlled experiment. Such experimental data limits the range of permissible probability assignments, though the method for arriving at such a range remains controversial. Suppose we find a diversity of probability assignments within the permissible range. How could we find a single probability assignment to use as a summary of the divergent assignments? One

51

R. Hilpinen (Ed.), Rationality in Science. 51–61.

method would be by voting, and another would be to average the probability assignments. Such a procedure might, however, not be the best
method available. One reason is that either such procedure would give
equal weight to the opinions of each scientist. We might have information about the scientists in the group that leads us to conclude that
certain scientists should be given greater weight than others. Some are
more expert, have a better record of prediction, better training, and so
forth. This information, which I have called social information, is germane to computing a reasonable summary of the information concerning the probability of *B* or *R*.

The problem is to assign the appropriate weight to each expert. In the
assignment of weights and probabilities, I am a Bayesian subjectivist, at
least of a moderate variety. This implies that there is no simple rule an
individual *must* use to arrive at such an assignment. In some instances it
will be reasonable for a person to equate the probability of a hypothesis
with some observed frequency. In assigning a weight to someone, it
might be reasonable for a person to equate the comparative weight that
a person is assigned with his comparative success rate at predictions of
the kind in question. But such an equation may not be always rationally
compelling. There will be cases in which one has not *counted* to ascertain frequencies, either pertaining to a hypothesis or pertaining to the
success of a person in prediction. One may have great amount of non-
mathematical data about a subject or person that is germane but not
directly computable as a frequency. Defenders of subjective probability
have explained how one might arrive at probability assignments without
frequency data in terms of a sufficiently complete articulation of preferences. Moreover, even in cases in which one has some frequency information, one may prefer to rely, at least in part, on nonstatistical
information. For example, the training of a young scientist may be an
important factor in deciding how much weight to give to his opinion
when one has no exact statistical information about how frequently
scientists with that training are correct. Such nonstatistical information
may lead us to give some weight to his opinion even if he was wrong in
the single prediction we have observed. Statistical data are important for
arriving at a probability assignment, but one may reasonably choose not
to *equate* probabilities or weights with success frequencies.

How, then, are we to ascertain the appropriate weight to assign to
each member of a group? Starting as subjectivists, we might have each
member assign a weight to every other member of the group. We might

ask each member of the group to divide a unit vote among all the members of the group, including himself, on the assumption that these weights will subsequently be used to determine the appropriate weight to be assigned to the opinion of each person concerning the probability of B or R. Each might assign some arbitrary value L to the person to be given the least weight, the value M to the person to be given the most weight, and scale others in between with gambles. No one method seems to be rationally compelling, however. So, we leave the matter of dividing the unit vote among the members of the group to the sagacity of the individual, requiring only that the weight be assigned in the interest of arriving at the truth or the best estimate of the truth in the probability assignment ultimately reached. Egoistic interest in having one's own interest prevail is disallowed. Each individual is obliged to assign weight solely in terms of the comparative merits of the individuals in the group.

Let the weight w_{ij} be the weight that person i assigns to person j. Once the weights are assigned, there are two possible outcomes. One is that each member of the group assigns the same weight to any specified member, j, of the group. In short, there might be unanimity about what weight to assign to each member. We then let w_j represent the consensual weight unanimously assigned to j by the group. In this case, the reasonable probability assignment for B would be $\sum_{i=1}^{n} w_i p_i(B)$, where w_i is the consensual weight assigned to i and p_i is the probability assignment of i.

There is one special case in which unanimity of weighting would be rationally mandatory for arriving at the truth. Suppose that there is a population and we are attempting to ascertain what percentage of the population has a certain characteristic. Each member of the group observes a subpopulation in such a way that the subpopulations do not share any common members and the set of subpopulations observed by members of group exhausts the total population. In this case, to arrive at the correct answer about the population, each person should equate his estimate about the frequency of the characteristic in the population with what he finds in his subpopulation, and the weights assigned to members of the group should equal the proportion of the total population contained in the subpopulation he observes. The summation principle just given would then yield the correct answer for the total population.

The normal expectation when individuals assign weights to others would be that different individuals would assign different weights to a specified individual. This means that in addition to a diversity of probability assignments for the hypothesis in question, we would have a diversity of weights for individuals. The weights are relevant information for arriving at a reasonable summary or consensual probability assignment. We need, therefore, some method for amalgamating or aggregating this information to find a reasonable summary.

The simplest answer would be to find an appropriate weight to assign to an individual by averaging the various weights that members assign to that individual and using the average weight thus obtained to derive the summary probability assignment as in the first case described above. This strategy has the following defect. Suppose that some member i of the group assigns a very high weight to a person j. As a result, j has a high average weight, and his probability assignment strongly influences the summary probability assignment. Imagine that everyone else in the group assigns a very low weight to i. That is, others in the group, perhaps with the exception of i and j who might have a small mutual admiration society, think that i is very unreliable. Would it be reasonable to let the very high weight which the unreliable i assigns to j strongly influence the outcome? It would not seem reasonable to proceed in this way.

There is a much better method which, given one restriction, will enable us to obtain a summary probability assignment. Suppose that, in the absence of a consensus about what weight to assign to each individual, we ask each individual to average as in the first case above – but using the weights that the individual assigns to members of the group to arrive at an improved probability assignment. That is, starting with the initial probability assignment p_i^0 of i in the initial state 0, an individual arrives at a new probability p_i^1 as follows: $p_i^1(B) = \sum_{j=1}^n w_{ij} p_j^0(B)$. He arrives at a new probability assignment by taking the weighted average of the probability assignments of members of the group using the weights that he assigned for the averaging.

When we ask an individual to proceed in this manner, we can justify our directive on the grounds of consistency. If the individual refuses to average in this way, we can tell him that this is equivalent to assigning himself a weight of unity and everyone else a weight of zero and averaging. If, in fact, he assigns others some positive weight, then the state 1 probability assignment is an improved probability assignment in terms of the weights that he assigned. To refuse to average is to discount *his* information about others to zero. He may object to this argument, but assume he does not. Suppose each member in the group moves from the initial state to a state 1 probability assignment. There may still be disagreement about the probability of B. We then suggest that each individual further improve his probability assignment by taking a weighted average of the state one probability assignments. This amounts to adding one to the superscripts in the equation just given. The justification for this improvement is the same as that given above; to refuse to average is equivalent to now discounting the information one has about others to zero. From stage two we obtain further improvement by averaging to reach stage three, and so on.

The averaging process just described has some interesting mathematical properties. The most important is that it may converge as the improvements occur. We can be sure the process will converge when positive weight is communicated from every member of the group to every other member of the group. Positive weight is communicated from i to j when there is a series of members of the group, the first being i and the last j, such that each member assigns positive weight to the next member in the group. Notice that this means that it is possible that positive respect is communicated from each member to every other member when each person in the group gives positive weight to the next member in a circle of the members. To say that process converges is to say that as the members proceed from state to state the probability assignment of each member converges toward a common summary or consensual probability assignment for the group, $p_c(B)$. Thus, the method yields the sort of summary probability assignment we sought.

The mathematics of this process yields an interesting insight into what the averaging achieves. The original weights give us a matrix

$$A = \begin{bmatrix} w_{11} & w_{12} & w_{13} & \cdots & w_{1n} \\ w_{21} & w_{22} & w_{23} & \cdots & w_{2n} \\ \hline w_{n1} & w_{n2} & w_{n3} & \cdots & w_{nn} \end{bmatrix}$$

and the original probability assignments give us a vector

$$p^0 = \begin{bmatrix} p_1^0(B) \\ p_2^0(B) \\ \hline p_n^0(B) \end{bmatrix}$$

The averaging that takes us from the initial state to state one yields the result equal to multiplying A by p^0, that is Ap^0. State two is then $A(Ap^0)$, that is, A^2p^0. State n is A^np^0.

The foregoing mathematical observation tells us that the process described above is equivalent to a process of finding the appropriate weight to assign each person, that being our original objective. For given the communication of positive respect described above, raising the exponent on the matrix will bring convergence of a special kind. There will be a weight w_j such that all the weights in column j will converge toward w_j as the exponent is raised. (There is, however, no guarantee that all the weights will equal w_j in a finite number of states.) What this means is that the averaging procedure, though aimed at improving an individual's probability assignment, amounts to a process of finding an appropriate or consensual weight for each person and then using those weights to take a weighted average of the original probability assignment.

It is important to notice that this method, which we have applied to a question in science, could equally well be applied to a question of science policy. Instead of taking the weighted average of probability assignments, one might have taken the weighted average of a utility assignment over possible outcomes of policy. Hence, for example, if the question was whether to build a nuclear power plant or a plant that uses conventional fuel, one would assign utilities to the possible outcomes of these actions to calculate the expected value of the various courses of action. In order to calculate the expected utility of a course of action, one must assign some value or utility to a possible outcome of an action and a probability to obtaining that outcome. The probability assignment might be reached by the method described above so that a group decision is based on the summary of all the information the group has about what outcomes will result from alternative actions. This leaves us with the assignment of utilities. Suppose there is disagreement about what utility to assign to an outcome C. We may then again reflect on how much weight to give to the members of the group. Each member could ask himself how he would divide a unit vote among the members of the group. Those weights can then be used to find some summary or consensual utility assignment by the same method that was used to find a consensual probability assignment. Moreover, the justification for the use of the method to find such a utility assignment is the same as in the probability case.

It is important to notice that the weights assigned for probabilities might be different from the weights assigned for utilities even though they are assigned by the same persons to the same persons. The question a person might ask himself to arrive at an assignment of weights might be different in the two cases. To arrive at weights in the probability case, a person might ask himself how he would divide a unit vote among the various members of a group if the votes were to be used to choose the best person to estimate probability, the best guide to truth. To arrive at weights in the utility case, a person might ask himself how he would divide a unit vote among the various members of the group if the votes were used to choose a person to estimate utility, the best guide to satisfactory policy. It is quite possible that we might think that a person is a good guide as to what the consequences of an action will be but a poor person to choose a policy on the basis of that information. For example, we might think that a man is an excellent engineer who is quite reliable as to the probable consequences of constructing a nuclear power plant, but we might also think that the person is rather indifferent to the fate of the people who might be affected. So, we assign him a high weight in the probability case and a low weight in the utility case.

The use of the consensual utilities and the consensual probabilities may be used to calculate expected utility by specifying states of nature S_1, S_2 and so on to S_n, and by letting O_{ij} be the outcome resulting from performing action A_i when outcome state S_j prevails. The formula for the consensual expected utility of A_i, $e_c(A_i)$, based on the consensual probability assignment, p_c, and consensual utility assignment, u_c, is $E_c(A_i) = \sum_{j=1}^n p_c(S_j/A_i)u_c(O_{ij})$. The appropriate social policy would then be the one that has a maximum of expected value. If instead of being interested in a question of policy, one is interested in some question within science, for example, a question of whether it is reasonable to accept some theory or hypothesis, then one might reasonably restrict the choice of a utility function to capture the intellectual objectives that are appropriate to science. If we allow that there are two outcomes of accepting a theory, or hypothesis, that one accepts what is true or that one accepts what is false, and we let u_t and u_f represent the intellectual value of those outcomes, then the expected intellectual value of accepting H, $eI(H) = p_c(H)u_t(H) + p_c(\sim H)u_f(H)$.[2] It is controversial how the intellectual utilities should be assigned, and it is possible that the best method again would be to find some summary or consensual assignment. The consensual assignment can yield the basis for computing what it is reasonable to accept in science and what policy it is reasonable to adopt for the utilization of science.

OBJECTIONS AND THE MODIFIED METHOD

There is one fundamental difficulty that warrants introducing a modification of the model. Let us reconsider the probability case. Suppose that each person has averaged once, thus moving from the initial state zero to state one. Then suppose that i is considering what weight to give to person j for the next round of averaging. As he considers the weight to assign to j, he notes the state one probability assignment of j, $p_j^1(B)$. That is equal, i notes, to $\sum_{k=1}^{n} w_{jk} p_k^0(B)$. Therefore, when i assigns a weight to j at this point, w_{ij}, and multiplies that by $p_j^1(B)$, then he is multiplying w_{ij} times w_{jk} and multiplying that product times $p_j^0(B)$. So at this level, the weight i assigns to j is a weight that i gives to j as a weighter of other members in the group.

The problem is that the weight that i gives to j as a judge of solar astronomy might differ from the weight that i gives to j as a judge of solar astronomers. Moreover, when we move to state two, the weight that i gives to j will amount to the weight that i gives to j as a weighter of weighters of other members of the group. And, again, just as the weight that i gives to j as a judge of astronomy might differ from the weight that i gives to j as a judge of solar astronomers, so the weight that i gives to j as a judge of solar astronomers might differ from the weight that i gives to j as a judge of those who judge solar astronomers.

Thus, rather than assume that the weights assigned at each stage should be the same, we must allow that the series of weights w_{ij}^1, w_{ij}^2, and so forth that i gives to j to arrive at state one, state two, and so forth, might differ. This means that rather than multiply one matrix A repeatedly by a probability vector to obtain a consensual probability assignment, we have a sequence of matrices A_1, A_2, and so forth, so that our products are $A_2 A_1$, $A_3(A_2 A_1)$ and so forth. The question is whether such multiplication yields convergence.

My collaborator, Carl Wagner, has obtained the following answers to this question.[3] Let us call a matrix that has identical rows, that is, one that is such that all the value weights in a given column are the same, a consensus matrix. Then we obtain the following results. If A_i is a consensus matrix, then $A_j A_i = A_i$. This means that if there is consensus at the level of A_1, then the higher level weighting will preserve consensus. Moreover, if A_j is a consensus matrix, then $A_j A_i$ is a consensus matrix. This means that if there is consensus at any level, consensus will be preserved.

Moreover, even if the matrices are such that no one of them is a consensus matrix, the sequence of products A_1, $A_2 A_1$, $A_3(A_2 A_1)$, ... may converge toward a consensus matrix. If the matrices A_1, A_2, ... converge toward a matrix A in which positive respect is communicated from each member to every other member, then the sequence of products will converge toward a consensus matrix. Let us assume that all the matrices satisfy the condition that positive respect is communicated in the requisite manner. If after a certain level k, all the higher order matrices are the same, then the sequences of matrices will converge in the appropriate manner so that the sequence of products of the matrices will converge toward a consensus matrix. It is, I believe, plausible to suppose that convergence of this type will result, because after a certain level it is natural to suppose that no new information will be available. Thus, for example, if i assigns j a weight of 0.3 as a weighter of weighters of others in the group, it is reasonable to suppose that if i asks himself what weight to assign to j at the next level as a weighter of weighters of weighters of members of the group, he will have no new information and again assign j a weight of 0.3 as he did at the previous level. In short, it is reasonable to suppose that no new information would be available beyond a certain level and that no change in weighting would result beyond that level. Therefore, the sequence of products of matrices would converge toward a consensus matrix.

One other objection to the averaging of weights concerns the independence of the first level weights and the probabilities, as well as the independence of weights at different levels. One might be able to contemplate these matters in such a way as to compartmentalize the various assignments. Wagner has proposed a method for accomplishing this.[4] To arrive at the initial probability assignment, members of the group circulate anonymous position papers about astronomy, and probability assignments are made on the basis of these papers without revealing the authors. If there is dissensus, the authors of the position papers are revealed and first level weights are assigned. To assist this process anonymous papers are exchanged concerning the expertise of members of the group as solar astronomers. If there is no consensus about what weights to assign at the first level, the authors of the second set of position papers are identified and second level weights are assigned. Again anonymous position papers are exchanged concerning the expertise of members of group as judges of solar astronomers. And so forth. This process insures the independence of one level from the next through the anonymity of the position papers. Moreover, it is reasonable to suppose that at some level the anonymous position papers would contain no new social information. From that level on, the weights assigned should not change thus insuring convergence toward consensual weights and a consensual probability assignment.

EMPIRICAL SPECULATIONS

If we suppose that the simple method is, in some cases, a plausible idealization of actual consensus, some interesting predictions would result. First of all, note that one might have two groups of people that are not linked by communication of positive respect. The combined population has, therefore, no consensus. As members of one group become better acquainted with members of the other group, a consensus for the combined group may emerge as a consequence. This consensus may be at variance with the consensus within one of the original groups. From their perspective a catastrophic or revolutionary change may appear to have occurred without any basis in background information. That, however, would be an illusion. The change is a consequence of the communication of social information about the expertise of members of the combined group

Finally, the model presents a model of the impact of a group on an individual and of an individual on a group. At the early stages of life or career, an individual can be expected to have the social or consensual beliefs and values inculcated or imposed. The individual has little impact on the theories or values inculcated, because little weight is given to his opinion. The consensus at that time, in terms of our model, would not be influenced much by his opinion. As the individual gains knowledge and expertise the situation will shift. He may then be given some weight, perhaps considerable weight, by his peers and thereby influence consensual opinion. Indeed, he may even dominate such opinion. This seems to represent a model of the interaction between an individual and society in terms of beliefs and values. If we suppose that consensual opinion dominates the individual as some social scientists have suggested, we may also note that the social opinion may be strongly influenced by an individual.

I should like to conclude with a problem. It is part of the lore of science that the person who does not conform to consensual authority often makes the greater contribution. Similar reflections apply in the moral or practical sphere. The individualist and iconoclast is a hero. On our model, such individuals appear unreasonable in that they ignore the amalgamation of social information. There are two explanations of this that are compatible with the model. One is that there is a consensus that one should not always conform to the consensus. Thus, one may become a consensual iconoclast. The other is that the individual may sometimes have information that social information is unreliable because no one is very reliable on the issue in question. It is worth noting, moreover, that there may be a consensus to the effect that social information is unreliable. Perhaps the resolution is to note that there appears to be a consensus that when it is the consensus that the social opinion on a subject is not well informed, the individual should ignore social opinion. When the consensus is that consensus is not reliable in some domain, ignoring the latter consensus is a reasonable consensual policy.

University of Arizona

NOTES

[1] Lehrer, Keith, 'When Rational Disagreement is Impossible', *Noũs*, Vol. 10, 1976, pp. 327–332; 'Social Information", *The Monist*, 60, No. 4, 1977, pp. 473–487; and 'Consensus and Comparison: A Theory of Social Rationality', in *Foundations and Applications of Decision Theory*, Vol. 1, C. A. Hooker, J. J. Leach, and E. F. McClennen, eds., D. Reidel Publ. Co., 1978, pp. 283–310.
[2] Still the best summary of such epistemic utility functions is Hilpinen, Risto, *Rules of Acceptance and Inductive Logic*, Acta Philosophica Fennica, 21, North-Holland Publ. Co., 1968. For my views on this topic see, Lehrer, Keith, 'Truth, Evidence, and Inference', *American Philosophical Quarterly*, 11, No. 2, pp. 79–92.
[3] Wagner, Carl, 'Consensus Through Respect: A Model of Rational Group Decisionmaking', *Philosophical Studies*, 34, pp. 335–349.

KUNO LORENZ

SCIENCE, A RATIONAL ENTERPRISE?

Some Remarks on the Consequences of Distinguishing Science
as a Way of Presentation and Science as a Way of Research

Our main concern has been to deal with various aspects of the concept of rationality. Specifically, the rationality of some scientific activities has been questioned, among them the case where doxastic attitudes are chosen, or where weights are assigned to experts in a group which aims at a consensus on some scientific question, or even where the guiding principles necessary for calling an activity 'scientific' are being adopted. More generally, we asked for criteria of rationality with reference to human behavior in verbal or non-verbal interactions or to human actions (i.e. intentional behavior) in general.

As an immediate consequence of these attempts to reach at an explicit determination of the (meta-rational) norm 'be rational' the question came up whether such a norm should be turned from a categorical one into a hypothetical one, valid only under conditions where rational procedures cannot conflict with what intuitively could be called progress due to imaginative ingenuity, i.e. a kind of "sound" irrationality counterbalancing the tendency towards a non-sensitive conservatism felt to be inherent in any kind of rationality. If, now, the principle hidden behind that progress is the old and venerated quest for truth the question arises whether rationality as a condition on means for truth as an end is really able to serve as a safe guide to truth. In other words, is there a chance to formulate conditions of truth for the results of scientific activities without recourse to the rationality of those activities: is it possible – to use a metaphor of Wittgenstein – to throw away the ladder after having climbed up on it to true results. Or, is the very claim for truth nothing but a reasoned and, hence, rational way to attain results.

In the following remarks I want to make use of the distinction between science as a way of presentation and science as a way of research which may throw some light on that seeming conflict among the two norms 'be rational' and 'be right'. The idea is that science as a way of presentation should be understood as a theory of meta-competence (the result of following the directive: be rational!) – a knowledge of the means to secure the truth of propositions about objects – whereas science as a way of research is a theory of object-competence (the result

R. Hilpinen (Ed.), Rationality in Science. 63–78.
Copyright © 1980 by D. Reidel Publishing Company.

of following the directive: be right!) – a knowledge of the objects through adequate representations of them. Object-competence can, hence, be acquired only in the presence of the respective objects; you have to deal with objects – appropriate speech acts included – to get acquainted with them. Meta-competence, on the other hand, works primarily in the absence of objects: it is just that tool on the level of signs which serves to counterbalance the lack of "knowledge by acquaintance" through substituting "knowledge by description" for it.[1]

And what has to be investigated is the interrelation of these "two ways" historically as well as systematically. I would like to start with my considerations a historical claim concerning the traditional two ways to cope with epistemic scepticism: rationalism and empiricism. Due to a misunderstanding on the part of the empiricist tradition up to the modern analytical philosophy of science, the ways of research have been erroneously treated as descriptions (i.e. on the meta-level) relative to given domains of objects, and due to a dual misunderstanding on the part of the rationalist tradition up to the modern constructive philosophy of science the ways of presentation have in turn been erroneously treated as constructions on the object-level relative to given domains of concepts. For further support of this claim I continue with a kind of historical sketch concerning the fate of rationalism and empiricism.[2] There are two main developments originating from well-known problems of epistemology in both of these philosophical positions which may be characterized in the following way: Out of rationalism emerges transcendentalism to secure a unique set-up of at least the natural sciences, mathematics included. We may call this the *a-priori*-method to set up mathematics and some fundamental parts of physics. Empiricism, on the other hand, gave way to evolutionarism, some kind of free choice principle to be used for starting, e.g. the sciences or any other human artefact. We may call this the observation-method to stick to what is at hand in a given moment. Now, it is common opinion to treat an epistemology of the first kind as the only way out of epistemic scepticism taken seriously, whereas an epistemology of the second kind bounds scepticism by some commonsense-relativism which implies to drop any reliance upon science as a substitute for religion concerning matters of fundamental world view.

This frame for dealing with the claims of scepticism hides that difference of presuppositions in the philosophy of science which I referred to as treating science as a way of presentation and treating it as a way of research. Of course, this difference takes up the old person-oriented *ars iudicandi* on the one hand and the matter-oriented *ars inveniendi* on the other hand, and I should add – just as a remark – that this Leibnizian difference of Analysis and Synthesis, as he alternatively calls these two "artes", is more general than Reichenbach's distinction between a context of justification and a context of discovery, since presentation (unlike Reichenbach's justification) as meta-conpetence is dependent on object-competence (otherwise the linguistic means cannot be safeguarded against loss of meaning), and research (unlike Reichenbach's discovery) as object-competence is dependent on meta-competence (otherwise mutual communication will lose any control of success or failure).[3]

Transcendentalism searches for justifiable presentations (being true theories of certain domains of objects) whereas evolutionarism represents a way of adequate research (being significant encounters with certain kinds of objects), such that in the first case we get well-founded sequences of propositions, yet in the second case a well-determined network of mutually related objects.

At once a further complication arises. To search for sequences of propositions is certainly not a purely linguistic matter. Instead of just giving a construction of certain objects on the language-level the crucial issue is to judge upon their "relation" to the object-level with the aim of securing their truth. Analogously, to represent a network of objects cannot be done on the object-level alone; representation is bound to rely on linguistic means with the aim to determine the objects by precise descriptions.

In order to avoid erroneous identifications, I have deliberately used here the terms 'search' and 'represent' to refer to activities within science as representation and science as research, respectively. For, certainly, there is second order research concerning presentations (e.g. within what is called 'science of science') and second order presentation concerning research (e.g. within the well-known 'logic of inquiry'), and neither should be identified with what I called 'search' and 'representation', respectively.

Search within presentations is search for *true* presentation, whereas representation within researches is representation of *significant* research. What is at stake from the purely linguistic point of view is the question of how the verification (and falsification) of formulae is interrelated with the signification of terms. Both questions, the justification of propositions and the constitution of objects, have to be answered separately yet dependent on each other and they must not be confounded.

If the question of constitution is falsely treated as belonging to the problem of justification it yields evolutionarism as a brand of radical empiricism.[4] And again, if the question of justification is falsely treated as belonging to the problem of constitution it yields transcendentalism as a brand of radical rationalism. To keep both questions within their proper bounds has a chance of success only if their mutual dependency is treated clearly and distinctly. This means especially to ask for methods to translate theories including states of theories – considering theory-change – into each other: In what sense may two synchronically or diachronically different theories have the same content – though saying different things of different entities.

The most prominent example for the radical empiricism – the empiricist misunderstanding – which arises when problems of constitution are treated as if they were problems of determination, i.e. as if they concerned investigations into the truth-conditions of assertions about the constituted objects, can be found in the evolutionary pragmatism of C. S. Peirce.[5] With slight modifications only, this is equally true of the "Analytische Wissenschaftstheorie" as it grew out of the logical empiricism of the Vienna Circle.

For example, the usual set-up of formalized theories has never been seriously questioned, i.e. the start with given domains of objects on the one hand and sets of predicates together with suitably chosen axioms about those objects on the other hand. There is a freedom of choice in both respects – e.g. phenomenalistic systems may compete with physicalistic systems, and preference for some set of primitive notions and principles should always be treated as a contingent fact itself – yet there is usually no hint as to how somebody can acquire a position enabling him actually to choose among alternatives. This again remains a historical and thus contingent fact.

In the converse case, the most prominent example for the radical rationalism – the rationalist misunderstanding – which arises when problems of justification are treated as if they were problems of constitution, i.e. as if they concerned investigations into the conditions of possible experience (how objects of experience have to behave in order to be accessible to knowledge) can, of course, be found in the transcendental idealism of I. Kant.[6] It is, therefore, not accidental when proponents of the "Konstruktive Wissenschaftstheorie" use arguments akin to Kantian ones to substantiate the claim that what is called 'protophysics' can serve as an a priori foundation for physics.

I will not go into further historical details now, but rather stress certain features of the discussion between the analytic and the constructive philosophy of science which are relevant for the epistemological issue in the sciences I am concerned with.

For convenience of presentation, I will start with the discussion of a thesis, which Harald Wohlrapp has convincingly defended few years ago:[7] The analytic philosophy of science on either of its three main stages, Carnap-Stegmüller's empiricism, Popper-Lakatos' rationalism, and Kuhn-Feyerabend's historicism, should – according to Wohlrapp's claim – essentially be understood as concerned with science as a way of research, whereas the constructive philosophy of science of the Erlanger Schule and of related positions, is basically concerned with science as a way of presentation.

As an important consequence, the difference of criteria for what shall be considered as science and, hence, as a rational enterprise can be stated. The criteria of science as research are essentially those of success, derived from the *actual* procedures of working scientists; necessary conditions are, e.g. the use of well-defined predicates, the reliance on the consistency of the set of non-derived sentences, reproducibility of operations *et alii*. On the other hand, the criteria of science as presentation follow conditions of acceptability and are in this sense "foundational"; they derive from *potential* procedures of scientists and can be characterized essentially by two principles: the principle of method (i.e. presentations work stepwise without "jumps" – a kind of completeness claim) and the principle of dialogue (i.e. presentations can be criticized, which means it can be questioned whether they fulfill the first principle).

If one proceeds this way and at the same time argues from a treatment of science in the first case as a fact and in the second case as a norm – how science is and how science should be – a certain kind of simplification occurs and affects the conclusion. The reason being simply that research cannot be described without recourse to decisions of what shall count as relevant and that presentation cannot be issued without using the content of the respective predicates. Therefore, it remains to be investigated whether the specific claims of failure and success Wohlrapp raised with respect to the two sets of criteria he discusses really carry conviction as they stand or whether failure and success rather depend on some further distinctions connected with the difference of research and presentation.

To repeat what I said in the beginning, I would like to claim that inasmuch as questions of constitution are confounded with questions of justification – or, to use the linguistic angle, questions of signification confounded with questions of verification – the criteria for science either as research or as presentation will lead to difficulties and eventually to failures. Success, therefore, in either case is dependent on a clarification of the interrelationship between answering on "what there is" and answering on "what is true", the ontological and the epistemological version of the question of how the two levels of objects and of signs ("world" and "language") separate within and unite into one domain of (scientific) language-games.

I have tried to show elsewhere[8] that this domain has to be understood as a domain of "preactions", beyond the action-act dualism as the prototype of the type-token division, and equally beyond the classical distinction of something given and something to do. The idea is simply to start with – from a later point of view, complex – objects which do not yet bear the differentiation between actor and action or between action and object of an action or result of an action. From that starting point to develop both ordinary language and the language of science as something on top of a more elementary and obviously fictitious language where only reference to such objects occurs, is already a piece of work during which most of what is treated later on as logical or ontological presuppositions of a language gets decided.

The difficulty is that in giving a description of this (re)construction the language of description, i.e. some standard natural language in use, is far more developed syntactically and semantically than the described language during the process of its construction.

Hence, in order to get an adequate account of the construction it is necessary to introduce certain devices which make sure that the description at any stage is not dependent on those features of the syntactic and semantic structure of the language of description which do not yet belong to the structure of the constructed language. For example, the difference of singular and general terms within the language of description should not be relevant for describing the initial stage of construction, where within a fictitious elementary language only reference to "preactions" occurs. Rather, there should exist an explicit step of introducing that very difference within the elementary language. And this is done by proposing to introduce the singular/general distinction on the elementary level of non-analysed actions as the distinction of *schema* and *actualisation*. These descriptive terms refer to the difference of "once", "once more", "once more again", ..., which is practically acquired in situations of repetitive imitation (= imitative repetition) with respect to any preaction.

It is obvious that such preactions are the pragmatic version of Strawson's "feature universals" in his essay *Individuals* (London 1959), at least with respect to their general aspect; Strawson forgets to include into his presentation their singular aspect as something on a par with the general aspect of preactions.[9]

Now, linguistic signs are the means which have developed gradually through our evolution to articulate the mutual dependency of schema and actualisation with respect to any preaction: it becomes possible to *say* which general object belongs to which singular object, i.e. under which concept a certain case falls or by which case a certain concept is fulfilled. Through language something singular acts as a *symbol* of something general, and, the other way round: through language something general acts as an *aspect* of something singular.[10]

Then, it is perhaps not any more offensive – by being liable to the pitfalls of a remake of the cartesian dualism – to say: the singular gives the empirical base, the general the rational design; the two cannot be separated; research starts with singularia, presentations with generalia. Since theories of both areas exist, hierarchies of theoreticity appear[11] and the situation, including the empirical/rational distinction, becomes confused.[12] Just as a remark, it is interesting to note that these preactions are the candidates which show the two aspects, the mental one and the physical one, in recent discussions about the theory of actions[13] only after the preactions have been developed into proper actions.

As I already mentioned in connection with Wohlrapp's paper, it serves still further confusion, if in the first case the methodological position of the analytic philosophy of science is characterized as being descriptive, whereas in the second case the methodological position of the constructive philosophy of science bears the label of being normative. Inasmuch as questions of constitution have consequences in terms of stipulations concerning the objects of scientific discussions the "definition" (I prefer the more general term 'introduction') of basic predicates about them is included – the insistence on the normative character of some fundamental part of science – let us call it 'proto-science' – is reasonably supported.

Similarly, questions of justification, concerning the context of these objects – the *use* of predicates, so to speak, not their *introduction* – give rise to descriptive aspects of any science.

Here again, it might be useful to recall that in general predicates within scientific languages are defined on given domains of objects – extensionally as certain classes of those objects. Hence, they cannot be treated as primary predicates, they are derivative with respect to the defining predicate for the domain of objects (i.e. the "substances" defined as the instantiations of that primary predicate). We, then, speak of properties, and the usual problems concern questions of whether properties hold of objects which obey certain other descriptions, and they never concern questions of elementary constitution.

Those constitutional questions – unless they are non-elementary, i.e. of second order, yielding domains of abstract objects (such processes are of course well-known and extensively treated everywhere) – occur on a language level which is itself of a theoretical nature only. It is the elementary level I referred to earlier and which can now be charac-terized as the one where terms are introduced, not the one where they are used as in ordinary non-scientific and usual scientific language. E.g. when you assert 'this leaf turns yellow', the constitution of objects like leaves (i.e. the introduction of the term 'leaf') is presupposed, whereas the constitution of objects like "yellows" (nominal use of the term!) is pushed to a second order level: Yellow is constructed as an abstract object, a "quality", turning the word 'yellow' from a non-primary general term – standing for a "characterizing universal" in the sense of Strawson[14] – into a singular term, a nominator, as I propose to call it. What remains in the case of the assertion in question is to judge upon the use of the terms 'leaf' and 'yellow' (or 'turn yellow') which by all standards is a question of true description.

Constructions remain within one language- (or object-)level, descrip-tions concern two consecutive levels. This difference is well-known, e.g. in logical theory, where formulae can either be constructed by formation-rules, or they can be described by means of suitably chosen predicates of a metalanguage (used for formalising the construction).

The fundamental notions are *partition* (of a whole into parts) in the case of constructions, and *attribution* (of a property to an object) in the case of descriptions. Mereology and set-theory provide the respective formalisations of these notions though their interrelation has by no means sufficiently clarified up to now.[15]

I hope these remarks give sufficient support for the claim that there is no simple correlation between science as research and stating what is (being the result of research) on the one hand, and between science as presentation and issuing what shall be (being the guarantee of presentation) on the other hand.

In either case the set-up of science is not only a question of justifying a corpus of sentences (used as constatives and/or as directives), but a question of introducing meaningful terms viewed at from different angles or better: distances, only, as object-competence from near by and as meta-competence from far away. Justification now includes the search for the truth (science as theory) together with the search for the good (science as praxis) as much as meaning exhibits both, aspects of (theoretical) signification and aspects of (practical) relevance.

According to John Rawls,[16] the good is taken over by justice, though the criticism of Tranøy shows[17] that the good is justice only in its social aspect; in its individual aspect it will have to be determined as freedom – socially granted through mercy according to Tranøy –, and there is no easy conciliation between these two aspects.

The corresponding two aspects of truth look somewhat different though well-known under the labels of truth as consensus and truth as correspondence. In its genuine epistemological context 'truth' refers to a qualified consensus among persons on some matter of common concern, e.g. in the way Peirce had indicated by equating truth with the ultimate opinion of the indefinite community of investigators in the long run.[18] In a more recent terminology it is said that such investigations, if they obey the qualifications (i.e. if they are conducted "rationally"), maximize *epistemic utility*.

Now, the usual correspondence-theory of truth as the alternative aspect to truth as (rational) consensus should rather be understood as a treatment of truth in an ontological context: truth gets equated with adequacy of linguistic representations of objects which, again in a more recent terminology, means to conduct investigations with the aim of optimizing *truth output*. We know reasonably well how to handle these two aspects of truth,[19] yet in the case of the good, i.e. the call for procedures to tackle the problem of justifying norms, we are on far less secure grounds. To comment on our discussion of possible conditions for the rationality of actions I should like to stress that the so called means-ends-rationality[20] takes care only of those cases where the means-ends-distinction is relevant for the action in question. Outside the area of technical (or instrumental) norms – norms count here as generalized directives towards an action-type – neither practical (or social) norms, where the required action is an "end in itself"[21] nor moral (or meta-ethical) norms, where rather attitudes towards actions (including forbearances) than actions themselves are of concern, can be treated in this way.

The antagonism of freedom and justice, the individual and the social aspect of the good, we have been observing is due, I claim, to that antagonism of social norms and moral norms which derives from a conflict between two second-order-ends: social norms aim at uniformity, moral norms protect diversity. The reason why instrumental norms which aim at uniformity, too, do not essentially, i.e. irrevocably interfere with moral norms lies in the fact that instrumental norms allow substitution of means for the same end.

Now, since language-norms, i.e. the use of a language, whether enforced by institutions or not, seem to me in their conceptual aspect to be cases of technical norms (here: social conventions), yet in its perceptual aspect (when signs are taken as objects in their own right) to be cases of moral norms (here: individual rights to choose one's own way to speak, e.g. the socio-historic background of the language-norms as conventions), one could at once understand why it makes sense to plead for a unified language for science and not e.g. for the arts, and why there is so much opposition to such a pledge since almost every scientific enterprise includes distinctive features of artistic idiosyncrasy.

This may furthermore be taken as a hint that it is again the difference of object-competence and meta-competence which not only governs the two ways of science, as research and as presentation, but also two ways to treat the rules on signs, especially language-norms, once perceptually – the way of art – and other conceptually – the way of science (of course not restricted to the natural sciences).

This brings me back to the main topic, how the difference and the interdependence between science as research and science as presentation can actually be characterized.

The relevant difference of research and presentation as against the simplified accounts I discussed above comes in when we look for the *support* of a scientific theory. As far as science is treated as a way of research, this support should derive *only* from the descriptive power of the theory relative to the singular objects (= perceptual cores) of research, though usually, in the analytic philosophy of science, due to the aforementioned lack of a clear separation between constitution and justification, this support is extended to include the explanatory power of the theory as well. It is common to use the term 'confirmation' (referring to non-elementary propositions in relation to relevant singular "data") in this respect, and this entails a confrontation with the – I dare say unsolvable – riddles and paradoxes of induction.[22] The simple reason for the claim that nothing beyond the descriptive power of a theory can be treated within science as research only, derives from the following considerations: The explanatory power of a theory refers to the kind of interrelations which obtain among the different propositions of the theory, especially to an assessment of the range of validity of fundamental principles like those of conservation in physics. Hence, an account of the explanatory power can be given only by judging upon the conceptual frame of the theory used for the argumentations in science as presentations.[23]

The argument in the alternate case run conversely: as far as science is treated as a way of presentation, the support of a scientific theory should derive *only* from its constructive power relative to the general objects (= conceptual frames) of presentation, though, usually, in the constructive philosophy of science, due to the same confoundation of constitution with justification, this support is extended to the regulator (and, hence, normative) power of the theory as well. The term 'approximation' (referring to elementary objects in relation to relevant general "idea[l]s") is in this connection occasionally used, and, as an equally disturbing consequence, it becomes necessary to handle the vexations of the is-ought-gap.[24] Here again, it is easy to see that in science as presentations any attempt to go beyond the constructive power of a theory and to judge upon its power to issue what shall be – unless this is treated as a second-order question only, i.e. as a question of what kind of scientific activity (rather than objects of activity) should exist – will need reference to the perceptual cores of science as research, e.g. to the encounters with singular objects in experimental situations.

What I should like to claim is that both, the explanatory and the regulatory (or normative) power of a scientific theory can be assessed properly and without bias only if the set-up of science cuts straight through the separating line of research and presentation. This means especially that on each level within the hierarchy of theories the link between constitutional and justificational questions – and that refers to the interdependence of constructive with descriptive procedures as well – must not be lost sight of. E.g. the constative metapredicate 'state' and the directive metapredicate 'bring about' on the kernel-sentence 'this is P' may be used to arrive at the linguistic representation of two speech-acts, a constative '[I] state that this is P' and the directive 'bring about that this is P!' such that 'this will be P' can be explained by the directive (i.e. a demonstration of a future state of affairs by means of a present volition) and that 'do P' can be normatively justified (= regulated) by the constative (i.e. a probation of a present imperative by means of a continuing ability). This shows at least by way of indication how the interrelations in question might be dealt with.

Hence, trying to determine an adequate meaning of the two central concepts (scientific) *explanation* and (scientific) *regulation* amounts to nothing less but a reassessment of whether and how a unified treatment of science is possible. For this purpose, the concept of unified science should no longer be understood in the original historical setting along with a developing analytic philosophy of science. In the light of the considerations just offered I claim that a unified approach to science, unless it falls victim to typical "Scheinprobleme" as the one concerning the possibility of induction or the one concerning a bridge over the is-ought-gap, has to consider with respect to activities both of research and of presentation. It has to develop a concept of science starting with a kind of unity of research and presentation, where the domain of (scientific) language-games uniting "world" and "language" in the sense I have outlined above becomes the result of the first step. These language-games of preactions together with their articulations can then be treated in their both aspects: matter-oriented (research, unfolding object-competence by introducing acquaintance) and person-oriented (presentation, unfolding meta-competence by using description).[25]

For visualization of what I am driving at, I may use an example of current dispute: the different approaches to (physical) geometry. Concentrating on the research aspect of physics, the (temporal) behavior of (physical) bodies relative to their spatial coordinates is judged with respect to quite general hypotheses concerning space-time-structure (explanation-bias!). The presentation aspect of physics, in the protophysics of the Erlanger Schule on the other hand, asks for a series of steps to introduce the fundamental concepts of geometry, chronometry, and hylometry in that order using "idealized" operations with (physical) bodies (regulation-bias!).

In the second case, what is done, is to provide meaningful terms – that they can be used successfully outside presentational questions is taken for granted. It is not surprising that certain propositions come out true a priori. In the first case, something completely different happens: here, propositions about given objects are tested to secure their validity, which means to treat them as empirically based. The introduction of the terms used is taken for granted inasmuch as presentational questions are considered to be a cura posterior. Though theoretical activities of supplying true descriptions govern science in its research aspect or, rather, because of them, the presentational necessities like introducing meaningful terms get neglected. And, conversely, the concern with practical operations to get adequate constructions of fundamental concepts for science in its presentation aspect seduces into thinking low of problems whether those concepts can effectively be used in research situations.

The real issue actually boils down to the question of whether the introduction of meaningful (geometric) terms like 'straight', 'n-times the length of' etc. can be treated as an *extension* of ordinary language about ordinary objects, serving better criteria of relevance according to further developed standards of significance and truth. For, if extendability fails, we are stuck in conceptual frames without prospects to satisfy them; and if presuppositions serve as substitute for explicit introductions, there is no chance to guarantee anything beyond the perceptual cores.

It is easy to see that the last two conditional sentences may serve as a modern and more refined version of Kant's famous dictum that concepts without intuitions are empty and intuitions without concepts are blind. The refinement consists in the introduction of hierarchies of theoreticity starting with (of course not uniquely determined) common sense experience phrased in everyday language.[26] The domains of objects of scientific discourse have to be arranged on levels of ascending and descending order without any chance to argue definitely for a universal "lowest" level – e.g. of elementary particles – sufficient for arbitrary future theories.

The usual arguments between protophysicists and "deutero-physicists" – if I may coin that term for the moment – using coordinate-systems with a spatial or even a spatio-temporal metric are beside the point as long as the problem of introducing a "metric" is exempt from a truly mutual discussion. What can be introduced rather than merely postulated on the basis of elementary common human experience (still ambiguous relative to the singular – general bifurcation) will lend itself to the discrimination of (empirical) actualisations from (rational) schemata for any preaction.

Universität des Saarlandes,
Saarbrücken

NOTES

[1] Cf. for the terms together with the idea B. Russell, 'Knowledge by Acquaintance and Knowledge by Description', *Proc. Arist. Soc. N.S.* 11 (1910/11); for further relating this distinction to the distinction of presence and absence of objects (expounded during the work in the research project on 'Wissenschaftssprache versus Umgangssprache. Probleme des Aufbaus einer Wissenschaftssprache in Literatur- und Kunstwissenschaft' conducted by D. Gerhardus and K. Lorenz, sponsored by the Deutsche Forschungsgemeinschaft from fall 1977 to spring 1980), the two steps of a primary and a secondary dialogue-situation in K. Lorenz, *Elemente de Sprachkritik. Eine Alternative zum Dogmatismus und Skeptizismus in der Analytischen Philosophie*, Frankfurt 1970, become relavent.

[2] Parts of the following exposition derive from a further elaboration of parts of the paper by the author, 'The Concept of Science. Some Remarks on the Methodological Issue 'Construction' versus 'Description' in the Philosophy of Science', in *Transcendental Arguments and Science* (P. Bieri, R. P. Horstmann, L. Krüger, eds.), Dordrecht 1979.

[3] Cf. R. Reichenbach, Experience and Prediction, Chicago/London 1938, esp. ch. I, and compare with the context of Leibniz's terms as expounded e.g. in H. Hermes, 'Die ars inveniendi und die ars iudicandi', *Studia Leibnitiana Suppl.* III, Wiesbaden 1969.

[4] A term used by W. James for his version of pragmatism, which is exactly in line with the claim just made, cf. the collection of essays in *The Philosophy of William James* (W. R. Corti, ed.), Hamburg 1976.

[5] Of course, the radical empiricism of W. James may be included, too, since this issue can be dealt with quite independently from the dispute between James and Peirce on the meaning of the term 'pragmatism'; cf. for support, e.g., Peirce's argumentation against first intuitions to secure cognition in 'Questions Concerning Certain Faculties Claimed for Man', C. S. Peirce, *Collected Papers* I–VI (ed. Ch. Hartshorne and P. Weiss), Cambridge, Mass. 1931–35, 5.213 ff.

[6] This derives from the fact that Kant never disputes the reality of knowledge, i.e. of Newtonian physics, but tries to clarify the conditions of its possibility; cf. the relevant exposition in the last chapter (§15) of J. Mittelstraß, *Neuzeit und Aufklärung. Studien zur Entstehung der neuzeitlichen Wissenschaft und Philosophie*, Berlin/ New York 1970.

[7] H. Wohlrapp, 'Analytischer versus konstruktiver Wissenschaftsbegriff', revised version, in *Konstruktionen versus Positionen. Beiträge zur Diskussion um die Konstruktive Wissenschaftstheorie* (ed. with an Introduction by K. Lorenz), Berlin/New York 1979, Band II (Allgemeine Wissenschaftstheorie), 348–377.

[8] The latest account in my introduction to the reprint of R. Gätschenberger, *Zeichen, die Fundamente des Wissens*, Stuttgart 1977; at the same place attempts to relate this approach with ideas of the symbolic interactionism as developed by G. H. Mead and of the genetic epistemology by J. Piaget.

[9] Under the headline 'Property and Substance' being the terms for repeatable and non-repeatable entities respectively, the same issue is at stake when R. M. Rorty in his paper on 'The Subjectivist Principle and the Linguistic Turn' (in *Alfred North Whitehead. Essays on his Philosophy* (G. L. Kline, ed.), Englewood Cliffs, N.J. 1963) discusses – and refutes – the attempts of A. N. Whitehead to evade well-known epistemological dilemmas deriving from the singular-general dichotomy, if this dichotomy is correlated in a straightforward way, i.e. without using linguistic analysis, with the body-mind dualism.

[10] This idea is basic, already, to Peirce's treatment of the sign-process in the general framework of his theory of categories; cf. Lecture IV on 'The Reality of Thirdness' (*Collected Papers* 5.93–5.119) among the 'Lectures of Pragmatism' (the singular being existent only, is an object of knowledge through that which is real, i.e. the universal).

[11] Representative the treatment in the last chapter of W. V. O. Quine, *Word and Object*, Cambridge, Mass. 1960, §56 (Semantic Ascent).

[12] Cf. the sophisticated treatment of the "empirical core" (= empirical content) of a theory via Sneed's criteria of theoreticity as expounded e.g. in W. Stegmüller, *Probleme und Resultate der Wissenschaftstheorie und Analytischen Philosophie, Band II (Theorie und Erfahrung)*, 2. Halbband (Theorienstrukturen und Theoriendynamik), Kap VIII, Heidelberg/New York 1973.

[13] Representative the treatment of D. Davidson in his paper 'Actions, Reasons and Causes", in *The Journal of Philosophy* 60 (1963); the incorporation of both the mental and the physical aspect into the concept of action which is a natural consequence of the approach via preactions as proposed here makes it possible for Davidson to defend the much disputed Aristotelian claim of a causal connection between reason and action.

[14] Cf. P. F. Strawson, *Individuals*, London 1959.

[15] Cf. the paper by the author; 'On the Relation Between the Partition of a Whole into Parts and the Attribution of Properties to an Object', in *Studia Logica* 36 (1977).

[16] John Rawls, *A Theory of Justice*, London/Oxford 1973.

[17] Cf. K. E. Tranøy's paper in this volume pp. 191–202.

[18] Cf. one of the earliest versions in 'How to Make our Ideas Clear' (*Collected Papers* 5.388–5.410): "The opinion which is fated to be ultimately agreed to by all who investigate, is what we mean by the truth, and the object represented in this opinion is the real." (5.407).

[19] The pragmatic concept of truth is being developed in the game-theoretic approach of dialogic logic – cf. the collection of essays in P. Lorenzen and K. Lorenz, *Dialogische Logik,,* Darmstadt 1978–; the semantic concept of truth uses the well-known model-theoretic approach initiated by A. Tarski in his classical paper 'The Concept of Truth in Formalized Languages', cf. A. Tarski, *Logic, Semantics, Metamathematics. Papers from 1923 to 1938*, Oxford 1956, 152–278.

[20] Cf. the exposition in the paper by Lars Bergström in this volume, pp. 1–11.

[21] This refers back to the Aristotelian distinction of actions as means for some outside end (making – ποίησις) and as ends in themselves (doing – πρᾶξις), discussed e.g. in *Eth. Nic.* A and Z4.

[22] Cf. the discussion of the interrelation between explanation and induction in C. G. Hempel, *Aspekte wissenschaftlicher Erklärung*, Berlin/New York 1977 (German translation of a revised version of the last chapter of C. G. Hempel, *Aspects of Scientific Explanation and Other Essays in the Philosophy of Science*, New York 1965).

²³ Cf. for comparison the related remarks on the difference between descriptive and explanatory adequacy of a theory, here: of linguistics, in N. Chomsky, *Aspects of the Theory of Syntax*, Cambridge, Mass. 1965, Chap. I (Methodological Preliminaries).

²⁴ On the alleged interdependence of a constructive and a normative approach to science cf. F. Kambartel and J. Mittelstraß, (eds.), *Zum normativen Fundament der Wissenschaft*, Frankfurt 1973, espec. the essays by J. Mittelstraß ('Das praktische Fundament der Wissenschaft und die Aufgabe der Philosophie'), P. Janich ('Eindeutigkeit, Konsistenz und methodische Ordnung: Normative versus deskriptive Wissenschaftstheorie zur Physik') and O. Schwemmer ('Grundlagen einer normativen Ethik'). The independence of 'is' and 'ought' is usually taken for granted, formalized as non-validity of $\Delta! A \prec A$ in deontic logic (cf. P. Lorenzen, *Normative Logic and Ethics*, Mannheim 1969, p. 70f); and attempts to question the is-ought-gap get criticized even by other opponents of the analytic approach, cf. e.g. K.-O. Apel, who in his detailed discussion 'Sprechakttheorie und Begründung ethischer Normen' (in *Konstruktionen versus Positionen. Beiträge zur Diskussion um die Konstruktive Wissenschaftstheorie* (ed., with an Introduction by K. Lorenz), Berlin/New York 1979) takes pains to refute J. R. Searle's claims that there exist nontrivial logical relations among is- and ought-sentences. For further discussion cf. W. D. Hudson (ed.), *The Is-Ought-Question*, London 1969.

²⁵ For further constructions in order to reach the usual level of syntactic differentiations, cf. K. Lorenz, 'Words and Sentences. A Pragmatic Approach to the Introduction of Syntactic Categories', in *Communication and Cognition* 9 (Gent 1976).

²⁶ The grades of theoreticity which serve as a kind of measure for the distance to common-sense-experience (relative to some natural language system) get discussed by W. V. O. Quine, "Grades of Theoreticity', in *Experience and Theory* (L. Foster and J. W. Swanson, eds.), Cambridge/Mass. 1970.

MIHAILO MARKOVIĆ

SCIENTIFIC AND ETHICAL RATIONALITY

1

The problem is: how to make science more ethical and ethics more rational?

This sounds paradoxical: If science is the paradigm of rationality, as most people believe, why should it tend to become more ethical, why should it sacrifice its universality and neutrality and open its gates to moral value judgements which are notoriously irrational and relative to particular historical conditions? On the other hand, if there is something good in this irrationality and relativity of ethics, why destroy it by making ethics more rational and similar to science?

These paradoxes indicate a basic incompatibility between the way the problem was formulated and the way the terms "science", "ethics" and "rationality" are often interpreted. When by "science" we mean only a totality of systematically expounded, empirically confirmed, generally accepted knowledge – then it is indeed a paradigm of some kind of rationality. But science is also an institutionalized activity of research, motivated by all kinds of interests, purposes and utilities, having all kinds of consequences in its practical application. Much of it is extremely irrational, in some sense of that word.

On the other hand, if we mean by "ethics" and "morality" what the emotivist, analytical theory tells us: an area of culture where judgements are entirely, or to a large extent, mere ejaculations of individual or group feelings of approval or disapproval, then it would be pointless or even dangerous to infect science with such a spirit of complete relativism. But there are good grounds to hold that there are some universal constants in the morality of various societies and civilizations, (that there are some objective general human interests), and that some basic ethical principles may claim an objectivity and impartiality comparable with that of fundamental scientific assumptions.

Which leads us to the interpretation of the term "rationality". If rationality is reduced to the sphere of the intellect or, even more narrowly, of logic, then only a logic of moral language, not morality

R. Hilpinen (Ed.), Rationality in Science. 79–90.

itself, could be rational. However, rationality in a more general sense rules out the absurd, the arbitrary, the disintegrated, the particular deviating from a universal but *not* a *universal* human need of interest. The following conditions seem to be necessary for this broader concept of rationality:

(a)   All elements of the whole referred to as *rational* are consistent; contradictions must be resolved.
(b)   All that is rational, is regulated by explicit or implicit rules.
(c)   All rational thought or behavior is integrated and grounded on a set of basic assumptions.
(d)   Rationality involves a claim to universal validity within the given specifiable system of reference.

2

This concept of rationality embraces also rationality of practical action, the special case of which is ethical rationality. There is no need to claim that every historically known morality meets the four mentioned necessary requirements. Some certainly do, and there is nothing in the concept of morality that rules them out.

First, moral norms may be and often they are compatible with each other. But what happens if in a concrete situation there is a conflict of values? Each of them taken separately may express our genuine preferences "other conditions being equal". But when we have to choose between two preferences other conditions are no longer equal. Another basic type of moral conflict is one where we experience a conflict between abstract, conceptualized, verbal norms and an immediate, concrete moral intuition. The former has been internalized during the process of socialization. The latter may be the effect of a strong new moral influence, or the results of a process of critical assessment of the morality adopted in our youth. Moral dilemmas of this kind throw us into a problematic situation which a person who wants to stay both moral and rational cannot tolerate. If a morality is to be lived and sanity is to be preserved, then, whenever from a set of moral norms it follows that in a given type of situation we must act in a definite way, and we are immediately aware of the immorality of that type of action under given circumstances – then the contradiction must be resolved. Either we shall have to find out that in the network of abstract moral principles which have been learnt in the process of socialization and from which a controversial moral judgement was derived – there is at least one which has to be modified, generalized or plainly rejected. Or we must challenge and critically analyze our immediate, intuitive moral consciousness and find reasons to doubt its authenticity. We might have, after all been drunk, sick, depressed, in a state of shock, hypnotized or under powerful temporary impact of another stronger personality. We may be unable to resolve the contradiction and stay split for indefinite time. Our moral conduct would appear confused and irrational. In order to restore our integrity and rationality we will have to reconstruct the network of our ethical beliefs and to bridge the gap between the mediated and immediate moral consciousness.

Second, moral actions may be regulated by rules. We need not be aware of them, nor are we expected to always draw consequences from them before we act. But if morality of an action or intention is challenged we must be able to show how it follows from a set of rules – in the analogous way in which the justification of a scientific statement involves its derivation from a set of laws.

Third, the process of justification in both cases does not stop at particular rules or laws. These on their turn may be challenged by one who advocates a different scientific theory or different morality. One must be able to show how rules follow from some basic principles. A morality not less than a science could be an integrated whole.

Fourth, a relativist attitude based on an empiricist point of view is possible both in morality and in social sciences. But in the same way in which scientists and philosophers of science claim that truth is universal and the result of science universally valid, it is legitimate to hold that underlying enormous diversity of moral norms which hold under *different* social conditions there are some universally valid principles, which constitute the deep structure of the very possibility of existence and survival of any community. (Prohibition of incest is an uncontroversial example). From this point of view the defining characteristic of a moral action is that it follows a rule to which universal validity can be assigned. (Kant's "categorical imperative".)

In spite of all these similarities, scientific and ethical rationality are two different types of rationality. One is relevant to knowledge, where considerable harmless simplification and idealization is possible, the other to action in diverse, complex situations, where an immediate consequence of simplification may be human suffering. Therefore, it is easier to give account and lay down general criteria of the basic purpose of knowledge – the truth, then of the basic purpose of morality – the goodness. (The list of universal principles and laws of science is much longer and with fewer exceptions than that of ethics). Ethical rules are much less precisely formulated, if explicit at all. In the average, the moral experiences of different individuals in the same situation do not contain as many invariant, intersubjective elements as the sensory experience on which confirmation of scientific statements depends. The latter often involves quite simple quantitative observations of length, size, duration and coincidence with a standard, whereas the former is more qualitative, emotional and subjective.

Nevertheless, the difference is one of type, not of essence. Many among the greatest philosophers have refused to reduce rationality only to *theoretical reason* but have also discussed *practical reason*. In Aristotle's opinion *Phronesis*, practical wisdom, is one of the ways in which "the soul expresses truth".[1] Ordinary cleverness is "the power to perform those steps which are conducive to a goal we have set for ourselves". But "if the goal is base, cleverness is knavery".[2] Practical wisdom involves virtue – it is desirable in itself as much as theoretical wisdom. They differ in so far as theoretical wisdom comprises "the most precise and perfect kind of knowledge – *nous*, the true knowledge of basic principles, and *episteme*, all knowledge that logically follows from those principles". On the other hand, practical wisdom is concerned with things which can be different, with the changeable, possible, with human affairs, with action, and action has to do not only with universals but also with particulars.[3]

Kant makes a similar distinction between theoretical and practical reason (Vernunft). The former is the power of *knowledge a priori*. The latter determines our *will* to act, and lays down moral postulates.[4]

Aristotle considers practical wisdom inferior to theoretical wisdom, since it has essentially instrumental character "it makes us use the right *means*".[5] Kant on the contrary, establishes the primacy of the practical reason, since even those things which cannot in principle be theoretically known (such as freedom and immortality) have practical reality and must be presupposed in activity.

3

If there are strong theoretical and historical grounds to consider scientific and ethical rationality – two different varieties (theoretical and practical) of the same reason why, then, such strong attempts were made in some times and situations to identify them and in some different times and situations to completely separate them?

An emphasis on the identity between scientific and ethical rationality, as well as the tendency to blur the distinction between factual judgements and value judgements takes place in totalitarian societies when governing powers tend to increase their control over the life of all citizens using spiritual, ideological means. This was the case in the late mediaeval society. Religion provided the foundation for a unity between science and ethics. Very basic truths were allegedly transmitted to humankind by revelation. All theoretical knowledge and rules of behavior had to be derived from them – using formal logic as much as possible. In such a way science was subordinated to theology. The church fully controlled the daily practical life of the whole population.

Another example is bureaucratic socialist society. There ideology takes place of religion and the Party the place of the church. Again basic truths for both science and ethics come from the teachings of sacred classics. The emphasis is here much more on science – this is a society that aspires to an accelerated industrialization, urbanization and modernization of all life. But again the control over science is established in a peculiar way. The basic tenets of ideology have been proclaimed to be scientific and then projected back to science as obligatory truths; loyalty to these is, then, one of the basic criteria of the selection of scientific cadres. On the other hand, morality is completely fused (and confused) with ideology and politics. Morality is allegedly what serves the interests of the working class, which ultimately means the interests of the ruling elite which defines and represents them. The reputation of science is used to legitimize official ideology and morality, and to condemn any dissent as not only counter-revolutionary but also anti-scientific, irrational and immoral.

It turns out that the issues of factual- *versus* value-judgements and scientific- *versus* ethical-rationality are not nearly such academic issues as might at first appear.

Under conditions of religious or ideological hegemony it is historically justified to emphasize the difference between propositions of fact and those of value, between the character of science and that of ethics. When the ruling powers demand a monolithic unity of all segments of culture and the unity of official culture and behavior – any emphasis on pluralism, on diversity, on disunity is an act of emancipation. Max Weber was right in insisting that, in the conditions of restricted freedom of scientific research and teaching, the principles of ethical neutrality may save the honour and dignity of a scholar by allowing him to disengage from the immoral goals of the ruling circles. Renaissance resistance to the absolute power of catholic theology begins with the doctrine of two separate kinds of truths. In one of the relatively early expressions of resistance to Stalinism, Kolakowski wrote the essay "In Praise of Inconsistency". Inconsistency allows some degree of tolerance, it is "consciously sustained reserve of uncertainty, a permanent feeling of possible personal error, or if not that, than of the possibility that one's antagonist is right".[6]

Struggle for inconsistency, for the separation of the ethical and scientific makes sense when the evil, the irrational is overwhelmingly strong and dominating. Then it is the struggle for at least some little room where reason may survive.

There are other times and situations, however, when unreason and alienated power could be and should be stopped. Then it would be more appropriate to struggle for rationality in both science and ethics and emphasize their unity; science should not remain purely descriptive and ethics purely emotivist.

4

This is nowadays the situation in Western liberal societies.

Emphasis on the neutrality of science with respect to all values, which had a liberating effect in the struggle against ideological obscurantism becomes quite a serious limitation at a time when the practical application of science obviously leads to increasingly wide-spread and dangerous abuses. In an almost nihilistic absence of any historical orientation, in a situation of a spiritual vacuum, a value-free science helps to promote an ideology of consumption, of belief in material wealth and brute power over nature and other human beings. It was inevitable, sooner or later, to reach a stage, when the power indiscriminately generated by science will suffice to completely jeopardize some natural harmonies indispensable for life, and to create means for man's total self-destruction. Being deprived of any deeper humanist purpose, lacking any long-range projection (which is not a mere extrapolation of the existing social arrangements) value-free science becomes an easy pray and accomplice of the official ideology.

This is a different type of ideology than the one that characterizes totalitarian societies. Since they guarantee a considerable degree of political freedom and individual civil rights, liberalist societies are relatively stable in spite of all occasional turbulence caused by economic and cultural inequality. Ruling classes here need not aspire at full control over the bodies and souls of their citizen in every sphere of social life. The basic purposes here, as in every older organism, is to conserve the *status quo*. This can be intelligently achieved by encouraging particular interests to freely clash and cancel each other out, and by preventing a strong legitimate public emancipatory interest to emerge. Therefore liberalism had to change profoundly. When it still was a revolutionary philosophy it incorporated a number of normative concepts, such as *natural order, natural rights of man, progress, equality, sovereinity of the people, right to revolution*. These were tools of a critical evaluation of society. Two centuries later much of former idealism gave way to scepticism. Science renounces any critical evaluation and hands it over to politics. Ethics renounces any objectivity and relegates it to science, where it will be reduced to mere description and analysis. Value-free Science now plays a system supportive role in several ways.

First, "pure", positive knowledge can be interpreted and used in a way most convenient to those who are in control of political power and wealth. Properly selected positive scientific information is ideal raw material for an intelligent ideological propaganda.

Second, missing any critical spirit science inevitably takes the present state of society as the *normal* state, the parts of which are well *integrated*, each one having a definite *function*. All deviation from this order is dysfunctional, deviant, *pathological*. Thus "alleged neutrality" actually turns into apology.

Third, science that is concerned about means but not about ends, and that renounces any responsibility for its practical application has generated a technology that is dehumanizing, and a mode of production that is dangerously wasteful and self-destructive. On the other hand, present-day enormously accumulated alienated power in the form of the state, the army, the corporation, rests increasingly on the achievements of value-free science and technology.

5

It follows, then, that no matter how much individual scholars might internalize a "false consciousness" that their scientific endeavours were "value-free" – these are not. Their work serves a purpose, an interest, no matter how external and unconscious. A mathematician working on a paper for the Navy, a professor teaching science of government in Harvard are only obvious examples although even they might cherish illusions that they are carriers of pure knowledge – as Bertold Brecht's *der Denkende*, who must not fight, must not perish and must survive power because he transmits knowledge to coming generations.

The real problem is: what kind of purpose, of *telos* gives meaning to an inquiry; is it particular or general, is it autonomous or heteronomous, is it rationally justifiable or not.

Science becomes more ethical when it is guided by universal human interests, and when these are accepted autonomously.

Certain universal ethical values are implicit in the very concept of objectivity that constitutes the basic features of scientific method. Geiger was right when he insisted that there is a link between scholarly skills (*Fachkönnen*) and scholarly conscience (*Fachgewissen*). Objectivity presupposes a basic honesty in the application of the professional norms of research; merciless elimination of any personal vested interest, a cooperative spirit in the whole process of symbolic activity (without which true communication would be impossible); readiness to give priority to truth over group loyalty; or personal utility; freedom from religious and ideological intolerance. The objectivity of scientific research is contingent upon certain social conditions, and these, in turn depend in the implementation of a whole series of other values, such as the *openness* of a society toward the rest of the world, a general atmosphere of political and cultural *tolerance*, the *free* flow of information (which includes freedom of self-expression, of discussion, of travel, of studying any scientifically interesting problem), the *autonomy* of science from other social spheres, especially from politics, a social climate that favours a critical spirit and especially anti-authoritarian attitudes. On the contrary, any barriers to communication, any attack on integrity and dignity of scholars, any pressure to provide scientific arguments for the national cause of for the increase of profitability of goods, any ideological hostility toward rival philosophical approaches and methodological orientations, any monopoly of power that imposes control and censorship over scientific research and publication, and tends to promote loyal supporters into scientific authorities – greatly reduce objectivity and lead to a general deterioration of scientific work.[7]

There are such interests which indeed have *universal* human character, for example: survival of human race, increase of freedom and dignity, promotion of creative work and social cooperation, improving conditions of life, learning, transfer of culture and meaningful communication. Far from jeopardizing objectivity of science, these universal interests are objective by the very fact of their universality. Furthermore they constitute fundamental presuppositions of any objective scientific research.

It is conceivable, however, that in individual scholar for all kinds of reasons does not really accept the universal validity of those purposes and interests. If he is made to conform, he loses his autonomy and stops being moral. On the other hand he may internalize some entirely different goals, aristocratic, Nietzschean or Buddhist ones. He will be a moral person if he is indeed convinced that what motivates him deserves to be a general rule of human behavior, consequently that all social life should really be organized according to his basic goals.

This brings us to the problem of increasing rationality of ethics, especially ethics that is involved in science.

6

How to make the ethics of a critical science more rational?

It is possible to give up strict separation of science and ethics, agree that indeed most scientific research, especially in social sciences, is guided by hidden or explicit values but, then, insist that these are purely emotional and subjective. Thus one would accept that there is an implicit ethics in science but deny that there is any ethical rationality. These elements of ethics would suffer from relativism and would therefore constitute a foreign substance in the body of science.

Such a view could be challenged on two grounds. An action is ethical only if it is guided by a rule that is recognized as a universal maxim of behavior. This discards relativism. Rival judgements and actions may, as a matter of principle, be tolerated and respected but their ethical validity will be denied. Furthermore a value may appear as external therefore as irrational with respect to a closed rational system S_1. However, the irrational, the emotional itself may become an object of study within a broader rational system S_2 which embraces S_1 as a subsystem. When we explain why the apparently irrational motive of judgement had to take place in view of a number of its determinants, and when we justify it with respect to some higher-level values then what was an irrational element with respect to S_1 becomes a rational element within S_2. Thus

purposes and motives become psychological causal factors, goals turn
out to be merely means. But there is no reason to stop searching for ever
deeper causes and ever more basic goals.

Most scientists do not do that. When one works on a new bomb, or
develops an economic theory in order to make governmental policies
more efficient, or comes up with technological innovations in order to
make production more profitable, or brainwashes the patient in order to
make him adjust to social environment, he does not believe that it has
anything to do with scientific rationality to ask the questions. Is the
bomb worth while? Is it rational to make governmental policies more
efficient? Why should production be more profitable? Why should one
adjust to a sick social environment? These issues are considered external
to the body of science. They come from the politician, the business man,
the military – they indeed remain irrational. But their rationality may be
examined. Then it would be established that either they indeed are
ethical, i.e. may be derived from some basic moral principles or they
entirely lack any moral justification, and are merely the expression of
momentary, pragmatic particular interests.

An interesting case is when two opposite, incompatible utilites have
been derived from two different sets of moral principles. Suppose we
have two opponents who agree that science is almost never value-free
and at least tacitly presupposes some value judgements; that these value
judgements may be rationally examined within broader value systems;
and that these values indeed have ethical character, since they allow
universalization. Nevertheless, the two opponents disagree in their value
judgements since their basic assumptions are incompatible.

Is any further discussion at all possible and meaningful? Is the situation here entirely different than in science?

It is unlike but not entirely incongruous.

In both – basic assumptions cannot be theoretically justified.

In both inconsistency would be tolerated only exceptionally, as a
lesser evil.

In both cases capacity to embrace the rival system as its own special
case would demonstrate its superiority.

Both would refer to experience for confirmation. That is where the
essential difference lies. On the one hand, well controlled sensory observation secured in a scientific experiment is simpler than moral experience, and the logical connection between experience and the statement
to be tested is much clearer in a scientific than in an ethical dialogue.
On the other hand, the latter does not allow so much play with logical

possibilities as the former. One of the most frustrating features of scientific debates, and especially in abstract sciences, is that one may almost always escape refutation of his theory by playing conventionalist game: by introducing ever more conceivable possibilities and by turning ever more propositions of fact into conventions, i.e. definitions of concepts.

Game is also possible in ethics. For example, when asked whether he accepts a consequence of his moral assumptions our opponent might say he does, although in fact he does not. But then one must live his ethics – whereas one does not live his symbolic logic. We are not expected to go around and make statements of the kind "If snow is black then two times two is four". But we are expected to live our moral principles. And if we seriously assert that "life is not worth living", that "human society ought to be wiped out", that, other conditions being equal, "destruction is better than creation", "slavery than freedom", "war than cooperation", "humiliation than dignity" – then the world has the right to expect from us to see how shall we live such philosophy, and shall we indeed begin by destroying our poor children and ourselves. The world cannot be expected to take seriously our talk about how one should live if we demonstrate the unreality of our talk in our own life.

On the other hand if we tend to survive and continue to live in human society and in history, and make efforts to develop science and culture – and within this context have a dialogue about scientific and ethical rationality, then the inherent, presupposed part of this *apriori context* are such human interests as: communication, social cooperation, creative activity, learning, increasing freedom.

To sum up:

Science may become more ethical by explicitly abandoning its illusory neutrality, by assuming responsibility for the practical use of its results, by supplementing rationality of *means* by rationality of *ends*, and by systematically developing a standpoint of cultural and social criticism.

Ethics brought thus into a close relation with science, (without however being confused with it), would become more rational by insisting on universalization of all its norms and judgements, by integrating all norms into a whole, and by grounding them in those demands of practical reason that made history of human society possible.

Serbian Academy of Sciences and Arts,
Belgrade

NOTES

[1] Aristotle, *Nicomachean Ethics*, book six, 1139^b

[2] *Ibid.* 1144^a

[3] *Ibid.* 1142^a

[4] Kant, *Kritik der praktischen Vernunft*, Vorrede

[5] Aristotle, *Op. cit.*, 1143^b ^{33}

[6] Kolakowski, *Toward a Marxist Humanism*, Grove Press, Inc. New York 1969, pp. 211–220

[7] See: Mihailo Marković, Ethics of a Critical Social Science, *International Review of Social Sciences*, **XXIV** (1972), No. 4.

WILLIAM NEWTON-SMITH

THE UNDERDETERMINATION OF
THEORY BY DATA

1. THE REALIST THESIS

Can there be theories which are underdetermined by all actual and
possible observations? That is, can there be logically incompatible but
empirically equivalent theories? In this paper which falls into three
sections it will be argued that a suitably refined version of this question
ought to be answered in the affirmative. In the first section of the paper
attention will be drawn to an insufficiently appreciated reason for being
interested in this question. For most recent discussion related to this
question has focused not on the question itself but on Quine's notorious
claim that as *all* theories are underdetermined, translation is indeterminate. If we consider the current fashion in the philosophy of science for
realism we will be able to discern other reasons for being interested in
the question. For as I shall argue giving an affirmative answer to the
question is not compatible with realism as it is standardly understood.
In section 2 I will consider the arguments that have been or might be
advanced for thinking that theories cannot be underdetermined. Having
rejected these arguments I will seek to establish that underdetermination
can arise by constructing examples. In the final section I developed two
different modifications the realist might make in his position in the face
of what will be called *the realist's dilemma*. While the reasons will be
given for favoring one of these responses, it will not be possible to
definitely adjudicate between them within the confines of the present
paper.

   With the exception of some distinguished authors (among whom are
Kuhn and Feyerabend) it is fashionable to espouse some form or other
of realism. Those who style themselves realists generally see their position as constituted by the following four ingredients. First, scientific
theories are either true or false and which a given theory is, it is in virtue
of how the world is. In this context a theory is to be thought of as the
deductive closure of a set of postulates and to speak the truth or falsity
of a theory is to talk of the truth of falsity of the conjunction of the
postulates. This will be called the *ontological ingredient* in realism. A

91

Hilpinen, R. (*Ed.*), *Rationality in Science.* 91–110.
Reprinted from the *Aristotelian Society Supplementary Volume* LII (1978), pp. 71–91, with
permission of The Aristotelian Society.
Copyright © 1978: *The Aristotelian Society.*

second, and arguably not independent ingredient, which will be called
the *causal ingredient* is the claim that if a theory is true, the theoretical
terms of the theory denote theoretical entities which are causally responsible for the observable phenomenon whose occurrence is evidence for
the theory. It would be of small comfort to learn that our theories are
indeed either true or false representations of the world if we were
precluded from being able to have rationale grounds for believing that
any theory is more likely to have one truth-value than the other. And
our quite insatiable desire to know manifests itself in the third ingredient which will be cited as the *epistemological ingredient*. This is the
claim that we can have warranted beliefs (at least in principle) concerning the truth-values of our theories.

Implicit in the characterization so far developed is the assumption
that the goal of the scientific enterprise is the discovery of true explanatory theories. The reasonableness of this goal as stated has seemed to
some to be called into question by the fact that the history of science is
a grave yard of falsified theories. Indeed, there seems to be evidence to
support the meta-induction that any theory is found to be not strictly
speaking true within, say, two hundred years of its being produced.
Unless one is willing to take the courageous line of arguing that things
are now or will be much different, the evidence points to the same fate
for our current theories. And, thus, it looks as though the exercise of the
epistemological power the realists assumes we possess will always end
in a negative verdict. This might incline us to wonder how rational it is
to pursue a goal when the evidence is that that goal will never be
realised. The realist counter-move is to argue that while all theories are
false, some are falser than others. That is, the historically generated
sequence of theories of a mature science may well be a sequence of false
theories but it is a sequence in which succeeding theories have greater
truth-content and less falsity content than their predecessors. This empirical thesis, which will be called the *thesis of convergence*, would, if
tenable, render it rational to pursue the goal of truth for we would at
least have some assurance that we were getting nearer the unobtainable
goal.

The thesis of convergence together with the ontological, causal and
epistemological ingredients constitutes a sort of minimal common factor
among the wide range of philosophers who in recent years have ad-
vocated a realist construal of scientific theories. These would include
Boyd, Harré, Hesse, Popper and Putnam among others.[1] This position

needs modifying, I shall argue, for as it stands it presupposes that the underdetermination of theory by data cannot arise.

2. THE UNDERDETERMINATION OF THEORY BY DATA

At this juncture we need to articulate with some greater clarity the thesis of the underdetermination of theory by data. Quine expresses the thesis as follows:

> Consider all the observation sentences of the language: all the occasional sentences that are suited for use in reporting observable events in the external world. Apply dates and positions to them in all combinations, without regard to whether observers were at the place and time. Some of these placed-time sentences will be true and the others false, by virtue simply of the observable though unobserved past and future events in the world. Now my point about physical theory is that physical theory is underdetermined even by all these truths. Theory can still vary though all possible observations be fixed. Physical theories can be at odds with each other and yet compatible with all possible data even in the broadest sense. In a word, they can be logically incompatible and empirically equivalent. This is a point on which I expect wide agreement, if only because the observational criteria of theoretical terms are commonly so flexible and fragmentary.[2]

That is, we have a case of underdetermination if for some subject matter we have two theories, T_1 and T_2, which are (1) incompatible and (2) both compatible with all actual and possible observations. It is essential to bear in mind the force of the reference to actual and *possible* observations. For the thesis is not the uncontentious claim that a situation could arise in which at some moment of time all observations, text, experiments made to date have de facto left two rival theories in the field. For in the context of the sort of underdetermination Quine has in mind the outcome of *any* possible observation would either support both theories equally or count equally against both theories. If there can be such situations, the question as to which (if either) of the theories is true or correct would be *empirically undecidable*.

According to Quine *all* theories are underdetermined by the data; that is, it is held that for any subject matter there are incompatible theories all of which fit the data equally well. I will refer to this thesis of Quine's as the *strong UT thesis*. By the *weak UT thesis* I will mean the thesis that there *can be* cases of underdetermination. It is not clear that Quine has any non-question-begging argument to support this strong contention. Indeed, he tends quite candidly to offer this as something on which general agreement can be expected. Not unexpectedly the response to this contention of Quine's has been one of skepticism. This is not to say that such skepticism has been well-grounded by the production of a multitude of forceful arguments. On the contrary, while there has been considerable discussion concerning Quine's claim that the indeterminancy of translation follows from the underdetermination thesis, this thesis itself has received scant attention. No doubt this is so because most writers have been inclined, not unreasonably, to retort that Quine has provided no reason to think the thesis tenable. In this paper I want to partially rectify this deficiency on Quine's behalf by providing a proof by example of the weak *UT* thesis.

One might wish to object that the thesis as formulated involves the untenable presupposition that there is some viable dichotomy between observational propositions and theoretical propositions. For Quine takes it that a pair of theories are incompatible yet empirically equivalent if and only if they agree on the distribution of truth-values over the set of observational propositions and disagree on the distribution of truth-values over the set of theoretical propositions. Certainly the assumption that there is a difference in kind between observational propositions and theoretical propositions is dubious. But presumably one who denies that there is such a dichotomy will nonetheless agree that propositions vary in the degree to which they are observational or theoretical. In this case there will be a spectrum ranging from the more observational to the more theoretical. That being so we ask of a pair of incompatible theories where along this spectrum they differ in the ascriptions of truth-values. Whether or not the theories are to be regarded as observationally equivalent and theoretically divergent will depend on where along the spectrum one is willing to say we have got theoretical enough. And as, *ex hypothesi*, we are dealing with a difference of degree and not a difference of kind this will be a matter for decision. In light of this we can reformulate the underdetermination thesis as follows: No matter where one fixes the point demarking the observational from the theoretical (so long as one does not put this point at either end of the spectrum) there can be (or, in the case of the strong version of *UT*, must be) theories which agree on the distribution of truth-values to propositions on the observational side of that point and disagree on the ascription of truth-values to some propositions on the theoretical side of that point.

Given that there is no natural bipartite division of the propositions of a scientific language into those that are observational and those that are theoretical, the claim that in a particular case we have underdetermination will have to be relativized to a particular division. Such a claim may be more or less interesting depending on the particular division made. The division implicit in the examples to be given is such as to render interesting the claim that these constitute cases of underdetermination. In what follows it is assumed that some division has been made and the claim that two theories are empirically equivalent is to be understood as the claim that the theories agree on the distribution of truth-values over those propositions deemed observational. The question arises as to whether counterfactual observational propositions should be included. For Putnam[3] has objected to Quine's exclusion of counterfactual observation propositions about what would have been the results of experiments that were not but might have been performed (an exclusion which Quine's horror of propositions would lead him to formulate in terms of sentences). This means that two theories which agreed on the truth-values of non-counterfactual observational propositions while disagreeing on the truth-value of counterfactual observational propositions would be regarded by Quine as empirically equivalent and by Putnam as empirically inequivalent. Except for the following comment this question will not be further discussed for my examples are empirically equivalent on Putnam's more stringent criterion. To see the problem which arises if observational counterfactuals are included let T_1 and T_2 be theories which are empirically equivalent in Quine's sense and let p be a counterfactual observation proposition which is assigned a different truth-value by these theories. Presumably if this is to establish empirical inequivalence for Putnam there must be a matter of fact at stake with regard to p. As will be seen later one response to underdetermination is to conclude that with regard to those propositions responsible for the underdetermination there is no matter of fact at stake. One who makes this response would not think there was a matter of fact at stake with regard to p if the only way to assess the truth-value of p was by reference to these propositions. In view of the fact that the viability of this response is one of the crucial issues about underdetermination it would seem reasonable that in assessing empirical equivalence those counterfactual observations whose truthvalue can only be assessed by reference to T_1 and T_2 should be excluded.

Before developing examples of underdetermination it will be fruitful to consider three strategies that have been deployed in arguing against the possibility of underdetermination. First, it is easy to see that on certain views of the nature of theories underdetermination could not arise. This would, for instance, follow trivially given a reductionist construal of theories which treated all theoretical propositions as translatable into observational propositions. And this result follows almost as trivially if one holds that what a theory really is is more perspicuously represented either by its Craig replacement or by its Ramsey replacement.[4] However, in the absence of convincing reasons for thinking of theories along the lines of Craig or Ramsey these results are of no particular significance. Indeed, one might argue that the possibility of underdetermination provides additional support for the claim that Craig and Ramsey's theories cast no particular light on the nature of theories.

A second strategy would involve arguing that considerations other than fit with observational data are relevant in deciding between incompatible theories. An argument of this form is suggested by the following passage from Swinburne's *Space and Time*:

Compatible with any finite set of phenomena there will always be an infinite number of possible laws, differing in respect of the predictions they make about unobserved phenomena. Between some of these ready experimental tests can be made, but experimental tests between others is less easy and between them we provisionally choose the simplest hypothesis. Evidence that a certain law is simpler than any other is not merely evidence that it is more convenient to hold that suggested law than any other, but evidence that the suggested law is true.[5]

One who thinks with Swinburne that simplicity is a guide to the truth might hope to decide between rival theories which are compatible with all actual and possible observations by comparing the theories as to simplicity. This will, however, be somewhat problematic in view of the notorious difficulties involved in producing a reasonable criterion of relative simplicity. And then assuming we had such a criterion there is no reason to assume that any pair of incompatible empirically equivalent theories will be such that one is simpler than the other. Indeed, I claim that on any reasonable criterion of relative simplicity the examples to be developed involve theories of equal simplicity. In any event there is no reason to think that simplicity as a guide to truth. To have evidence that it is one would have to have evidence that the world was simple. If we suppose that there are two rival theories which can only be chosen between on the grounds of simplicity, we will have already to have shown that the simpler is more likely to be true in order to have evidence that the world is simple in the requisite sense. Thus, it is not clear that there can be any non-question-begging argument to support a principle of simplicity in this context. This is not to say that there are not good reasons for preferring the simpler of two rival theories. For, after all, it may be easier to manipulate such theories. However, we can agree to this without agreeing that its being simpler makes it more likely to be true. Simplicity makes theories more likeable but not more likely to be true. I recognize that this is a contentious and unsubstantiated claim. That need not concern us here if I am correct in claiming that my sample theories do not differ with regard to simplicity.

The third strategy is exemplified in the following remark of Dummett's

Quine's argument for the indeterminancy in the stronger sense is based on the claim, over which, he says, he expects wide agreement, that there can be empirically equivalent but logically incompatible theories . . . but the claim is absurd, because there could be nothing to prevent our attributing the apparent incompatibility to equivocation.[6]

An adequate discussion of this objection would take us far beyond the confines of this present paper. However, we can at the very least show that it is not obvious that the move suggested by Dummett can always be made by constructing examples which it would be unreasonable to treat as cases of equivocation. And perhaps we can do slightly better than this. Suppose we have a pair of theories that appear incompatible but really constitute a case of equivocation. Let T_1 and T_2 be first-order formalizations of these theories (formalizations which stand to the theories as first-order formal arithmetic stands to arithmetic). If we have a case of equivocation, T_1 and T_2 will be mere notational variants of one another in the sense that there are definitional extensions of T_1 and T_2, T'_1 and T'_2, which are such that any theorem of T'_1 is a theorem of T'_2 and *vice versa*. So if the formalizations of two theories do not have common definitional extension satisfying these conditions, the theories do not constitute a case of equivocation. Given this is a sufficient condition of non-equivocation, the examples to be given are not equivocal.

The crucial assumption on which both examples turn is the assumption that time does not possess whatever structure or topology it does possess as a matter of necessity. That is, while we take time in fact to be linear, continuous, non-ending and non-beginning, there is no logical or conceptual absurdity or inconsistency in the supposition that it should have some other structure. This assumption that just what structure time possesses is something that cannot be discovered by *a priori* reflection but can only be ascertained by empirical investigation cannot be argued for within the confines of this paper. However the theories to be developed do lend support to this assumption.[7]

For the sake of the first example let us begin by thinking of time as linear, non-ending and non-beginning. Let us imagine that we have come to have evidence that a cyclical cosmological model best fits the universe. Thus we are lead to conjecture that the universe is a sort of giant accordion that always has been and always will be oscillating in and out. Let us suppose further that as the evidence comes it tends to support the bolder hypothesis (an hypothesis of greater content in Popper's sense) that the universe during any one period of oscillation is quite similar to the universe during any other period of oscillation. Finally someone casts caution to the winds and conjectures: The universe is at time t_0 in a state of type S_0. At some time t_1 in the future the universe will *again* be in the same type of state S_0. Further, following time t_1 the universe will run through a sequence of states qualitatively identical of states that it runs through between t_0 and t_1. And so on and so on it runs with boring repetition indefinitely into the future. And similarly, it has run with this lack of novelty for ever in the past. This conjecture, which will be called theory T_1, is the conjecture that *time is linear and history is precisely cyclical.*

Under the supposition that all the observations that can be made support theory T_1 we can construct a rival theory T_2 which will be equally supported by that data. This is the theory that *time is closed.* That is, time has a structure like that of a circle so that there is just *one* occurrence of each type of state. Each occurrence of a particular type of state, however, lies in both the past and the future. We can represent the contrast between theories T_1 and T_2 in the diagram given below—

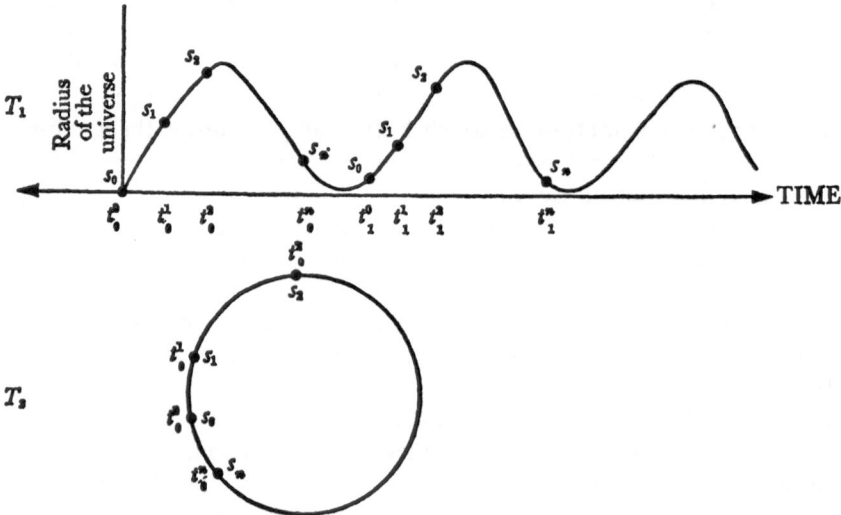

There are two points about these examples to which explicit reference should be made. First, in the story as told there is no possibility of distinguishing between, say, the times t_0^i and t_1^i by reference to differences in the mental states of observers present at these times. For the states of the universe including the mental states of conscious beings are qualitatively identical at these times. We can either suppose that someone present at t_0^i dies before t_1^i and is replaced by a qualitatively identical counter-part (who, depending on which theory we adopt, is or is not numerically identical to the counter-part person) or we can suppose that the person in question has a continued existence between t_0^i and t_1^i, in which case what he claims to remember at t_0^i is exactly what he claims to remember at t_1^i. Secondly, the picture associated with the closed time theory cannot be construed as a picture of the same time occurring again and again. To run one finger around the circle, so to speak, notionally counting the occurrences of t_0^0 is to embrace the incoherent notion that the same time could occur again and again. That is, it is to combine illegitimately the notions of closed time and of open time. In closed time each time is present but once. It is simply that what lies in the future lies in the past and *vice versa*. Admittedly this is, to say the very least, a most counter-intuitive picture of time. But as should be familiar from the character of many successful scientific theories of the twentieth century, intuitive, common-sensical judgments may be very bad guides to the truth both in the domain of cosmology and in the domain of quantum mechanics. Perhaps some of the counter-intuitive feeling is removed when it is realized that we could introduce with the help of a conventional stipulation an asymmetrical directed temporal relation which is locally transitive but globally non-transitive. Thus what lies in the local future does not lie in the local past and *vice versa*.

The theory T_2 will commend itself to one who – operating under the *initial* assumption that time is linear – reasons as follows. There is nothing that distinguishes the time t_1^0 in the future when the universe will be in state S_0 from the time t_0^0, the present time, at which the universe is in state S_0. For whatever is true of the time t_0^0 is true of the time t_1^0. Consequently the best conjecture to make is that the time t_0^0 is identical with the time t_1^0. In making this conjecture one is dropping the initial assumption that time is linear and replacing it by the assumption that time is closed.

T_1 and T_2 are clearly incompatible theories. However, the data that support T_1 support T_2 equally. The advocates of T_1 find that the data are exactly what would be expected if T_1 were true. The data in this case consist of support for the conjecture that the universe will in the future be in a state qualitatively identical with the current state. However, these data are exactly what the advocate of T_2 would expect given the truth of T_2. He too expects that there is in the future a state qualitatively identical to the present state. So, in the imagined context, the choice between theories T_1 and T_2 is empirically undecidable. The difference between the theories is shown by the fact that given T_1 the future time at which the universe will be in a state qualitatively identical with the present state is a time distinct from the present time and, on the other hand, given T_2 this future time at which the universe is in a state qualitatively identical with the state it is in at present is the same time as the present time.

It is important to note that the two theories in question ascribe non-isomorphic structures to time. That is, there is no one-one order preserving map between a structure having the topology of a closed curve and a structure having the topology of an open curve. It is this fact which guarantees that T_1 is not a mere notational variant of T_2.

It might be argued that the choice between these theories can be decided by non-empirical means. For instance, some have held that closed time is incoherent on the grounds that it is true in virtue of what we mean by 'past', 'present' and 'future' that no event can at the same time be past, present and future. In closed time when an event is present, it is then also past and future. However, even granting the claim about what we mean by these worlds this does not show closed time to be incoherent. What the argument in fact shows is that if we are to adequately characterize closed time we need to replace the simple tenses of ordinary language by more complex tenses.[8] It has, on the other hand, been argued[9] that non-empirical considerations could decide the issue in favor of closed time. For instance, some have taken it to be a necessary truth that if whatever is true at time t_0 is true at time t_1, t_1 is the same time as t_0 where the scope of the quantifier is restricted to propositions that can be expressed without reference to the times t_0 and t_1. If that were so, theory T_1 (open time, cyclical history) would be incoherent. However, it will not do in this context to rule out T_1 by claiming that the specialized form of the identity of indiscernibles given above is a necessary truth. For it is just this sort of situation (open time and cyclical history) that tests this claim. Indeed, I would argue that the principle cannot be necessarily true on the grounds that T_1 is an intelli-

gible theory. The two arguments above are but a sample of the considerations that might be introduced in an attempt to show that the choice
between T_1 and T_2 can be decided by non-empirical means, given that
T_1 and T_2 both fit the observable data. And the acceptability of this as a
genuine example of underdetermination is conditional on the assumption which cannot be justified within the confines of this paper that
there are no non-empirical factors which would decide the issue.

Before leaving this example I want to introduce an heuristic device for
thinking about this situation to which I will return below. To this end
consider two possible worlds A and B. We stipulate that in world A time
is linear and history precisely cyclical so that a theory of the type T_1 is
true of this world. We stipulate that the world B time is closed so that a
theory of type T_2 is true of world B. We assume also that the entire set
of states constituting world B is qualitatively identical with the sequence
of states in any one oscillation of world A. Imagine that you are to be
placed in one of these worlds. You will not be told which world it is.
The question then is – just what possible observation could you make to
ascertain whether you are in world A or in world B? My suggestion has
been that there is nothing you could do. While these are very different
kinds of worlds the question as to which world it is that you are in is
empirically undecidable.

The theories considered above do serve to make the point that the
choice between theories can be empirically undecidable. It must,
however, be admitted that these theories are not rich in explanatory
content nor are they nor have they ever been regarded as respectable
theories about the actual world. Consequently it is of interest to consider whether an empirically equivalent but logically incompatible rival
can be generated for a rich and sometime respectible theory such as
Newtonian mechanics. To see that this is possible we must first note
some salient features of Newtonian mechanics. If one develops a rigorous axiomatization of Newtonian mechanics it is necessary to postulate that space and time are continuous. That is, that the instants of time
have the order type of an interval of the real number line as do the
points of space along a given direction. This assumption is made as we
wish to represent the motion of a particle by a continuous function, the
position function, from real numbers (representing time) to triples of
real numbers (representing position in space). This guarantees that the
differential of the position function is well-defined and licenses us to
represent Newtonian mechanics by means of a family of differential

equations including for instance

$$F = \frac{m\mathrm{d}^2 x(t)}{\mathrm{d}t^2}$$

If on the contrary, we have assumed space and time to be merely dense and not continuous (*i.e.*, that the instants of time are isomorphic to an interval of rational numbers) we could not adopt Newtonian mechanics for the notion of a differential would not be applicable.

The assumptions of continuity mean that, restricting attention to point particles, we are assuming that the history of the motion of such a particle described in a particular frame of reference is to be represented by a set of ordered four-tuples of real numbers. However, when we directly measure duration or length we will never have occasion to use anything other than rational values, for any technique of measurement has some limit of accuracy and will be accurate only up to some finite number of decimal places. As a terminating decimal can be represented by a fraction we need only the rational numbers in assigning values to a directly measured parameter. There is just no technique one could follow which would lead one to record the results of an actual measurement by means of a non-terminating, non-repeating decimal, *i.e.*, by a non-rational real. In gathering evidence for Newtonian mechanics we will assign a rational value to the time at which the particle has some location which is assigned a triple of rational values. Putting these values into the equations will lead to a prediction about the position of the particle at some later time. The prediction may well generate non-rational values for the parameters. However, on checking the prediction we will assign only rational values to the parameters. If these values approximate to the predicted values this would be regarded as evidence for Newtonian mechanics.

The preceding considerations reveal a sense in which Newtonian mechanics goes beyond all actual and possible data. For while the data consist of assignments of rational values to the parameters in question, it is part of the theory that the parameters take real values which in some cases will be non-rational reals. This prompts the speculation that one could develop a rival theory with the same explanatory power which does not go beyond the data in this regard. Such a theory which will be called Notwen's mechanics can be developed. However, within the confines of this paper I will not be able to do more than provide a programmatic sketch of what would be involved in developing such a theory. We can think of Notwen as someone who shares Newton's general ideas about the relations between force, mass, acceleration *etc*. However, Notwen wishing to have a more parsimonious ontology assumes that space and time are dense but not continuous. Consequently he does not avail himself of differential equations in specifying the theory but characterizes the theory in terms of difference equations which unlike differential equations involve only operations that are closed in the rationals, Notwen's force law will be given as follows:

$$F = \frac{m[x(t + 2h) - 2x(t + h) + x(t)]}{h^2}$$

In effect Notwen's mechanics deals with average velocities and accelerations rather than instantaneous velocities and accelerations. The 'h' in the equation above represents a rational interval over which the averages are being taken. By taking h sufficiently small Notwen's equations can be made to approximate the Newtonian ones as closely as one likes. It is important that in specifying his theory, Notwen does not specify a particular value for h. Rather, his claim is that there is some rational value h for which the theory will fit all the data.

To have an adequate theory of mechanics one needs to do more than develop a system of difference equations which mirror the empirical predictions of the system of differential equations which constitute the core of Newtonian mechanics. One needs in addition, for instance, to assume a geometry for space. Notwen cannot avail himself of a differential geometry such as Euclidean geometry for that involves continuity assumptions. From Notwen's point of view it would be nice if one could develop a difference geometry analogous to Euclidean geometry. There are, however, certain technical problems involved in this project. For in *standard* measure theory it is not possible to give non-trivial measures for denumerably infinite sets. And under the density assumptions there will only be a denumerably infinite set of temporal instants and spatial points. I am inclined to regard this as a mathematical problem which it would be interesting to attempt to solve. However, even if this problem should prove intractable, it is not a decisive objection to Notwen. For Notwen could use the full range of mathematical techniques employed by Newton. In this case, in using an interval of the real number line in representing, for instance, the instants of time in an interval of time, Notwen would regard the non-rational reals as specifying ideal elements added for heuristic purposes and not as specifying actual instants of time. Only the rational reals would be regarded as identifying instants of time. That is, Notwen only affirms the existence of some of the items talked about in his theory. Talk of the other items is a convenient fictional device.

Notwen's theory with its postulation of merely dense space and time and Newton's theory with its postulation of continuous space and time are clearly incompatible. However the theories are empirically equivalent in the sense that an observation counts for (against) Newton if and only if it counts for (against) Notwen. Notwen and Newton will test their theories by measuring the values of the parameters and plugging these values into the equations to generate predictions. As was noted, the measured values with which they both begin will be represented by rational numbers. In a world in which Notwen's theory is successful a test of Notwen's theory will involve predictions of rational values for parameters which subsequent measurement supports. In this test Newton may predict the parameter to have a nearby non-rational value. However the subsequently measured rational value will be regarded as supporting the theory in virtue of being an approximation to the true value. On the other hand, if Newton's theory is borne out Notwen can find a value of h which is such that his theory is confirmed by the observations confirming Newton. In a similar vein we can see that observations which are disconfirming of one theory will be disconfirming of the other theory. Thus, the choice between these theories is an empirically undecidable matter.

It may be objected that I have not established that no empirical discovery could decide between these two pairs of theories on the grounds that the history of physics reveals the relevance of certain wider, more general empirical grounds than I have considered for holding one theory to have greater verisimilitude than a rival theory. To see how this objection might be developed consider a situation in which the actual available observational data underdetermine the choice between two rival theories. It might be that there is a more general theory of wider scope in which only one of these theories can be embedded. Given that the wider theory has some degree of empirical success it would be reasonable to opt for the theory compatible with that theory. Zahar has recently argued[10] that in 1905 the available data underdetermined the choice between Special Relativity and the Lorentz ether-drift theory but that only the Special Theory was or could be embedded in a gravitational theory (the General Theory of Relativity), and that this constitutes the empirical grounds for preferring Einstein to Lorentz. This situation might well arise in relation to my first example. For it might be that only one of the two rival theories is compatible with the best total physical theory we can devise where that theory as it turns out does not have, as far as we can tell, an empirically equivalent rival. While this is a possibility, there is no reason to assume *a priori* that the best total physical theory (if there be such a theory) will decide between the rival hypotheses or that there is a unique best total theory as opposed to two empirically equivalent rival total theories one of which favors the closed time hypothesis the other of which favors the open time hypothesis. Consequently I am inclined to concede that I have not established my example to be definitely a genuine case of underdetermination. However, it does serve to establish that there is no reason to assume that such a situation cannot arise. And thus the ball is put back into the court of one who insists that underdetermination is not possible.

In any event this style of objection does not seem to have force against my second example. For it would seem that a more general theory will decide in favor of, say, continuous time only if it involves some continuity postulates or other. And, using the devices I employed, one could construct an empirical equivalent rival of that theory which employed mere density assumptions. This theory would then decide in favor of the dense but not continuous space and time. So regress to more general theory in the case of this example will leave the example as a genuine case of underdetermination.

3. THE REALIST DILEMMA

Given that there can be cases of the underdetermination of theory by data, realism as characterized has to be rejected. For given that all actual and possible data relevant to some subject matter S falsify all theories for S except the incompatible theories T_1 and T_2 both of which fit the data equally well, the ontological ingredient in the realist position leads the realist to hold that there is something in virtue of which either T_1 is true or T_2 is true. However, *ex hypothesi*, nothing is going to count in favor of T_1 over T_2 and *vice versa*. In this context nothing could count as the evidence for thinking that T_1, is more likely to be true than false.

This is incompatible with the epistemological ingredient in the realist position which hold out the hope (at least in principle) of having warranted beliefs concerning the truth-value of our theories. To use one of our examples – the realist wants to hold both that the world is such that either time is closed or it is not and that we can come to have evidence concerning which it is. However, we have seen that in some contexts this is not possible.

At this juncture something has to give. One might try, on the one hand, to weaken the ontological ingredient. Or, on the other hand, one might try to weaken the epistemological ingredient. Initially anyone with realist sympathies inclines to respond to the dilemma by weakening the epistemological ingredient. One so inclined will insist that all scientific propositions have determinate truth-values and, while maintaining that in most cases we can (in principle) have warranted beliefs concerning the truth-values of our propositions, it will be conceded that this does not hold for the non-empty class of empirically undecidable propositions. The reason this response seems to be the most plausible, indeed to be the *only* plausible response, lies in the following two factors. First, if one has any sympathy with a realist position one will have adopted a 'correspondence' theory of truth, in the sense that a proposition is true or false in virtue of how the world is. Secondly, we tend to believe in the Law of Bivalence which amounts to the claim that any proposition is either true or false. We cannot abandon a 'correspondence' theory of truth without entirely extinguishing the spirit of realism. And given that there are at most two truth-values, and given that if a proposition has one truth value its negation has the other truth-value, a commitment to bivalence amounts to a commitment to the Law of the Excluded Middle (hereafter cited as LEM). Having a robust sense of common sense we don't see how we can abandon LEM. For it is, after all, one of the *immutables* in virtue of being a law of logic. (If you can't believe that what can you believe?) But these two ingredients basically constitute the ontological ingredient of realism. For example, by appeal to LEM we assert that either time is closed or it is not closed. And by appeal to the correspondence theory of truth we conclude that there is something about the world in virtue of which one or other of these alternatives is true. So the only way out of the dilemma seems to involve supposing that there are *facts* concerning which we can have no evidence. Now this response to which we seem driven is not entirely implausible. For surely it might be retorted that it was a piece of not inconsiderable arrogance in the first place on the part of the human intellect to assume that all there is to be known can be known by finite beings such as ourselves. This response, which I will call the *Ignorance Response*, involves maintaining that those propositions responsible for underdetermination are either true or false. It is conceded that with regard to these propositions we could not possibly have evidence concerning their truth-value. As such this response involves embracing the possibility of *inaccessible facts* – facts concerning whose obtaining or non-obtaining we could have no information.

Alternatively a realist might respond to the underdetermination of theories by restricting the scope of his realism in the following sense. Given a context in which some proposition P is empirically undecidable, the assumption that either P is true or P is false is withdrawn. As the realist holds that to be true (false) is to be true (false) in virtue of how the world is, this response involves dropping the assumption that there is something about the world in virtue of which P is true or something about the world in virtue of which P is false. That is, instead of supposing that there are inaccessible facts in virtue of which P is either true or false, we conclude that the world is simply indeterminate with respect to P. This response will be called the *arrogance response* for it amounts to holding that if we cannot know about something there is nothing to know about.

Consider the first example of an empirically undecidable proposition in light of these alternative responses. Many have a strong inclination to say of such a possible world that in it time either is closed or it is open – and that is that. Either the future occurrence of some state is a new and different occurrence of that state (*i.e.*, time is open) or it is numerically the same occurrence (*i.e.*, time is closed). Either time is such that it is like a closed curve or it is such that it is like an open curve. In making this response (the ignorance response) we are taking underdetermination as pointing only to our inability to have evidence concerning which of these possibilities actually obtains. That is, we have a case of inaccessible facts. One who makes the arrogance response will regard the heuristic device I introduced in the discussion of the possible world as inadmissible. The device in question involved us in imagining that we had been placed in one of a pair of possible worlds (in one of which time is closed and in one of which time is open). Such a move would be judged illegitimate by one inclined to make the arrogance response on the grounds that one is just not entitled to make these stipulations. For there is no determinate state of affairs which either obtains or does not obtain whose obtaining would make it true that time is closed. The set of facts constitutive of the one world, it would be claimed, *is the same as* the set of facts constitutive of the other world, and in that set of facts there is no fact answering to the proposition that time is closed and there is no fact answering to the proposition that time is open.

One who makes the arrogance response in the face of an empirically undecidable proposition P will not be willing to assert that either it is the case that P or it is not the case that P, and is thereby committed to denying LEM to have the status of a genuine law of logic. On the other hand, one who makes the ignorance response is likely to invoke the claim that LEM is a genuine truth of logic in attempting to justify his claim that there is a matter of fact at stake with regard to P – a matter of fact which is inaccessible. Consequently one inclined to the arrogance response may well wish to avail himself of the interesting arguments Dummett has explored for not asserting LEM.

The consequences of this line of argument strike some as implausible. For there are many cases in which our intuitive inclination is to assert a substitution instance of LEM (*e.g.*, Either a city will be built at the North Pole some day or a city will never be built at the North Pole) where, given the line of argument explored by Dummett, we would not be entitled to do this. However, there is a weaker strategy which might be deployed by someone to vindicate the arrogance response without embracing these apparently implausible consequences. To develop this strategy we need first to remember that we are dealing with propositions and contexts in which those propositions are empirically undecided in the sense that fixing the truth-value of all observation sentences leaves the truth-value of those propositions open. For instance, the first example of underdetermination provides such a proposition (that time is closed and history is unique) and such a context. One inclined, as I am, to the arrogance response sees no reason to admit in such a context the requisite inaccessible facts. Thus, without asserting that there can be no inaccessible facts, one asks of one making the ignorance response what his grounds are for asserting that there are inaccessible facts which either make it true that time is closed *etc.*, or make it true that time is open, *etc.*, (we are assuming that all other alternatives have been excluded so that the only way in which time can fail to be closed is if it is open).

It will not do for the advocate of the ignorance response to appeal to

LEM. For what is in question is just whether LEM holds for empirically undecidable propositions. Consequently it is not clear what could possibly count as a reason for thinking that there is a matter of fact (a matter of inaccessible fact) at stake here. For there is nothing that would be explained by the supposition that there are such facts. And that being so, we should prefer the ontologically weaker position which does not assert the existence of such facts. This means restricting the scope of LEM to exclude empirically undecidable propositions. If the under-determination of theory by data is a relatively rare phenomenon this will not mean a very extensive restriction. The limited scope of the restriction arises from the assumption that there is a matter of fact at stake with regard to any observational proposition. Given this assumption we can still assert, for example, that either a city will be built some day at the North Pole or a city will never be built at the North Pole. For a distribution of truth-values over the set of all observational pro-positions will determine the truth-value of 'There will be a city built some day at the North Pole'. One who denies this assumption will have to embrace a more extensive restriction on LEM.

To opt for the arrogance response means ceasing to regard empir-ically undecidable propositions as expressing hypotheses about the facts. Consequently the onus is on who so opts to give an account of the rôle those propositions play in the theories that contain them. One possibi-lity is that these propositions should be seen as serving to specify a mode of description or general framework within which the hypotheses about the facts are to be expressed. That is, for instance, one who asserts that time is continuous is not making a guess about the facts but is opting for a particular net for catching the facts. All the facts that there are can be expressed in terms of this framework; or, equally they can be expressed in terms of the rival framework based on the treatment of time as merely dense. This possibility will have to be explored elsewhere. My concern now is only to note the need of some such account.

My primary aim in this paper has been to produce reasons for think-ing that the weak *UT* thesis holds. A consequence of this is that the realist construal of scientific theories, as not uncommonly understood by philosophers of science, is untenable. For given *UT* the ontological and the epistemological ingredients in realism cannot be simultaneously satisfied.[11] Two different modifications (the ignorance and the arro-gance responses) that the realist might make in the face of this dilemma have been noted and a reason was given for favoring the arrogance response. Consequential modifications are required in the specification of the causal ingredient and in the thesis of convergence as these depend on the ontological and epistemological ingredients. The two modified forms of realism while quite different have much in common including the difficult and pressing problems of providing an analysis of verisimilitude and a defence of the convergence thesis, difficulties whose exploration will have to await another occasion.

Balliol College, Oxford

NOTES

[1] See in this regard M. Hesse, *The Structure of Scientific Inference* (Macmillan, 1974), 290; R. Harré, *The Philosophies of Science* (Oxford University Press, 1972), 90; R. Boyd, Realism, Underdetermination, and a Causal Theory of Evidence *Noûs* 1973; H. Putman, What is 'Realism'? *Proc. Aristot. Soc.* 1975/6, pp. 177–194.
[2] W. V. O. Quine, On the Reasons for the Indeterminacy of Translation *Journal of Philosophy* 1970, 179.
[3] Putnam, H. The Refutation of Conventionalism in his *Mind, Language and Reality* (Cambridge University Press, 1975), 180.
[4] J. English, Underdetermination: Craig and Ramsey *Journal of Philosophy* 1973, pp. 453–462.
[5] R. Swinburne, *Space and Time* (Macmillan, 1968), 51.
[6] M. Dummett, *Frege: The Philosophy of Language* (Duckworth, 1973), 617fn.
[7] This assumption is argued for in my *The Structure of Time* (Routledge, 1980).
[8] Ibid., Ch. 3.
[9] A. Grünbaum, *Philosophical Problems of Space and Time* (Reidel, 1973), 197.
[10] E. G. Zahar, Why did Einstein's Programme Supersede Lorentz's? *Brit. Journal Phil. Sci.* 1973.
[11] Boyd (*op cit.*) who has noted the tension between realism and underdetermination argues against the possibility of underdetermination. Reasons for rejecting his arguments are presented in my *The Rationality of Science* (Routledge, forthcoming).

ROLAND POSNER

TYPES OF DIALOGUE – THE USE OF MICROSTRUCTURES FOR THE CLASSIFICATION OF TEXTS

Texts are manifestations of language in use. In the attempt to describe the use of linguistic expressions in a text, the linguist must relate the communicative function of the text to its grammatical structure. The more comprehensive and less definite the text to be described, the more difficult the task of stating this relationship. Therefore, it is advisable, given the current status of linguistics and communications research, to begin with short texts that have a clearly definable communicative function.[1]

If we are successful in finding a correlation between grammatical structure and communicative function in small texts, we can then investigate the possible combinations of these small texts to larger ones. When a small text occurs as part of a larger text, its grammatical structure appears as a microstructure of the text as a whole, and its communicative function contributes to the communicative function of the complete text. The basic questions of the theory of texts – how does the communicative function of the partial texts determine the communicative function of the text as a whole; and how does the communicative function of the complete text affect the interpretation of its parts – can thus only be answered by investigating small texts.

In the following we consider in what manner the results of investigating small texts can be used to develop criteria for classifying longer texts. First we define the microstructures considered and present their properties in a series of theses. After that, we show how larger texts can be characterized on the basis of the microstructures contained in them. For reasons of space we will mainly be dealing with examples that the reader should generalize for himself. Instead of restricting ourselves to dialogue between constant speakers, one could extend the problem to include texts with changing interlocutors. And, instead of beginning with the microstructure of *comments*, one could bring in other types of specification.[2] The programmatic character of the following is intended to encourage the reader to engage in work of his own in the direction indicated.

R. Hilpinen (Ed.), Rationality in Science. 111–135.
Copyright © 1978 by Roland Posner

1. MICROSTRUCTURES

THESIS I. In uttering an independent sentence in standard situations, the speaker expresses *complex information* consisting of several pieces of information.

Even the simple sentence, "It is raining," conveys, beyond the factual information that rain is falling, temporal information that this is the case at the time when the sentence is uttered.[3]

THESIS II. In uttering a sentence, the speaker signals to the hearer what *communicative relevance* he places on the constituent parts of the information expressed.

What communicative relevance is, can best be illustrated by the following examples. Consider the testimony of two witnesses. Witness A says:[4]

(1) (a) Before the accused emptied the safe, he shot the watchman.

Witness B interrupts him protesting:

(1) (b) That's not true.

In this context (1b) explicitly means something like

(1) (b′) It is not true that the accused, before he emptied the safe, shot the watchman.

or

(1) (b″) The accused did not shoot the watchman before he emptied the safe.

In this exchange we are dealing with questions of criminal law. Witness A's testimony asserts, on the one hand, that the accused emptied the safe and, on the other hand, that he shot the watchman. Witness B's denial does not reformulate either of A's assertions explicitly. Nevertheless, any listener would conclude that only one of the two assertions is negated. The information that the accused emptied the safe remains uncontested.

Let us imagine another version of this controversy. Instead of saying:

(1) (a) Before the accused emptied the safe, he shot the watchman.

witness A says:

(2)   (a)   After the accused shot the watchman, he emptied the safe.

Witness B protests again by saying:

(2)   (b)   That's not true.

If, as before, we formulate explicitly what witness B has thereby expressed, then we have:

(2)   (b′)   It is not true that the accused, after he shot the watchman, emptied the safe.

or

(2)   (b″)   The accused did not empty the safe after he shot the watchman.

The sentences that witness A utters in these two scenarios, (1a) and (2a), are semantically equivalent; from either it follows that the accused shot the watchman and then emptied the safe. The commenting utterances of witness B are identical: "That's not true." However, by uttering this sentence, in one case B protests against the statement that the accused shot the watchman and in the other case that the accused emptied the safe. This effect cannot be due to the material relevance of the two statements; in criminal law the use of fire-arms is in any case of greater import than an instance of burglary. The fact that different information is denied in the two cases is rather a result of the speaker directing the attention of the listener differently; in version (1a, 1b) the communicative interest of both parties is focused primarily on the use of weapons, in version (2a, 2b) on the burglary. Thus the communicative relevance that the parties in this exchange assign to a piece of information must be taken to be independent of its material relevance.

Material relevance is a *semantic* concept; it indicates a relation between a given piece of information and a larger system of information, whatever the use of this information in communication. Communicative relevance is a *pragmatic* concept; it indicates a relation between the speaker and a given piece of information in a certain utterance. In contrast to semantic information, the pragmatic information of a sentence can be characterized by the fact that it is verified by uttering that sentence.[5]

The pragmatic information of a sentence can be further classified into two types: *locutionary* pragmatic information and *illocutionary* pragmatic information. When someone says, "Paul is coming," this sentence expresses the locutionary pragmatic information "the speaker presupposes that there is a unique person called Paul." And, on the other hand, when someone says, "Come!", this sentence expresses the illocutionary pragmatic information. "The speaker orders his addressee to do something."

Moreover, pragmatic information can be expressed implicitly as in the sentence "Come!". It can also be expressed explicitly as in the sentence "I hereby order you to come."

Placing communicative relevance on a piece of information in a sentence means assigning it locutionary pragmatic information.[6] Locutionary pragmatic information is normally expressed implicitly.

But let us return for a moment to the two courtroom scenarios. The only formal difference between them lies in the construction of the sentences (1a) and (2a). That is why we must conclude it is the phrasing of a sentence that signals which piece of information expressed has the greatest communicative relevance.[7]

Comparing the two courtroom scenarios not only points up the difference between material and communicative relevance, it also suggests how to discover what information in a given sentence has the greatest communicative relevance for speaker and hearer in standard situations; we need only add a comment of the type (1b) or (2b). Using such a procedure, we can define comment texts as follows: A comment text is a text constructed of two independent sentences uttered in immediate sequence with the second repeating a piece of information from the first as an embedded "that-" or "if-"clause. According to this proposal, the two sequences (1a, 1b) and (2a, 2b) are comment texts.

But we also want to be able to regard texts like (1a, 1b) and (2a, 2b), as well as (1a, 1b′) and (2a, 2b″) as comment texts. In order to do this, we must emancipate ourselves from surface syntax and refer to the semantic representations of the texts concerned:

DEFINITION i. If a text consists of two independent sentences uttered in immediate sequence and the second of these repeats a piece of information from the first as an argument of a sentence operator[8], then we call this text a *comment text*. The information expressed by means of the first sentence we call the *commentandum*, the information expressed by means of the second sentence we call the *comment*, and the information repeated in the second sentence as an argument of a sentence operator, we call the *commentatum*. The procedure of adding a comment to a commentandum we call *commenting*.

In principle, any piece of information in the commentandum sentence can be selected by the comment sentence as commentatum. It is only necessary to reformulate this information in the comment sentence. In (1a), instead of selecting the use of weapons it is also possible to comment on the information that the accused emptied the safe:

(1a)   Before the accused emptied the safe, he shot the watchman.
(3b)   It's not true that the accused emptied the safe.

or with reference to (2a), instead of selecting the burglary it is possible to comment on the information that the accused shot the watchman:

(2a)   After the accused shot the watchman, he emptied the safe.
(4b)   It's not true that the accused shot the watchman.

However, there is an important difference between the comment texts (1a, 1b) and (1a, 1b′), on the one hand, and (1a, 3b), on the other.

In (1a, 1b) and (1a, 1b′) it is the commentandum sentence that determines by its construction which part of its information appears as the commentatum in the comment; here the comment repeats the commentandum sentence without deletions or else refers back to it by means of a pro-sentence form.

In (1a, 3b), in contrast, it is the comment sentence that determines what information from the commentandum is commented on; here the comment repeats only one part of the commentandum as an embedded clause. The same difference holds between the comment texts (2a, 2b) and (2a, 2b′), on the one hand, and (2a, 4b), on the other.

Moreover, the selection of the commentatum can also be determined by the commenting operator in the comment sentence:

(2a)   After the accused shot the watchman, he emptied the safe.
(5b)   That was murder.

In (2a, 2b) the commenting operator is compatible with all pieces of information in the commentandum sentence, since it is equally possible to deny that someone shot a watchman as it is to deny that someone emptied a safe. In (2a, 5b), however, the commenting operator is not compatible with some pieces of information in the commentandum sentence because the sentence

(6)   That the accused emptied the safe was murder.

is unacceptable in standard situations. Thus, if the comment text (2a, 5b) must be paraphrased by

(2a)   After the accused shot the watchman, he emptied the safe.
(5b′)   That the accused shot the watchman was murder.

this is not due to the phrasing of the commentandum sentence but to particulars of the comment sentence.

These considerations militate for distinguishing two types of comments: *direct comments* and *indirect comments*.

DEFINITION ii. A *direct comment* is a comment that relies only on the phrasing of the commentandum sentence to select the information commented on. All other comments are called *indirect comments*.

Indirect comments influence the choice of the commentatum in that they only partially reformulate the commentandum sentence or being selectional restrictions to bear on it. Since in direct comments the selection of the commentatum is not influenced by particulars of the comment sentence, it can depend only on the communicative relevance of the pieces of information expressed by the commentandum sentence.

THESIS III. In a direct comment, *the communicatively most relevant information* of the commentandum sentence is made commentatum.

2. THE COMMENT TEST AND THE DELETION PROCEDURE

On the basis of thesis III it is possible to use direct commenting for the pragmatic analysis of sentences.[9] Consider, for example, the complex sentence (7a):

(7a)   When the accused claimed
          that the young man demanded still more money
          although the accused had given him ten dollars
          after he had been asked for it,
          the prosecutor objected that it was unlikely
          that such a person would be so unscrupulous
          that he would stoop to blackmail.

A direct comment in the form of a denial shows that the least subordinated clause in (7a) contains the communicatively most relevant piece of information:

(7)   (b)   That's not true.
(7)   (b′)   When ..., the prosecutor did not object that ....

If we now delete the clause with the most relevant information in (7a), then two information complexes remain:

(8)   (a)   The accused claimed that the young man demanded still more money although the accused had given him ten dollars after he had been asked for it.

(9)   (a)   It was unlikely that such a person would be so unscrupulous that he would stoop to blackmail.

A direct comment applied to (8a) yields:

(8)   (b)   That's not true.
(8)   (b′)   The accused did not claim that ....

Thus the information "the accused claimed something" is the most relevant information in (8a). If we delete it, we have:

(10)   (a)   The young man demanded still more money although the accused had given him ten dollars after he had been asked for it.

A direct comment applied to (10a) yields:

(10)   (b)   That's not true.
(10)   (b′)   The young man did not demand more money after ....

Thus, the information "the young man demanded still more money" is the most relevant information in (10a). If we delete it as well, we get:

(11)   (a)   The accused had given him ten dollars after he had been asked for it.

A direct comment applied to (11a) yields:

(11)   (b)   That's not true.
(11)   (b′)   The accused had not given him ten dollars after he had been asked for it.

Thus, the information "the accused had given him ten dollars" is the most relevant information in (11a). Let us delete it as well. What then remains is the information having the least communicative relevance in (7a):

(12)   (a) He asked him for money.

In the same manner it is possible to move through the information complex represented by (9a) to the clause with the least communicative relevance.

These results confirm to a certain extent the old claim – which is usually formulated normatively in rhetoric – that the "main clause" in the structure of a sentence expresses the "most important" information. The systematic application of the method of deletion on a large number of complex sentences with different types of constructions allows us to explicate and generalize this claim further:

THESIS IV: In unemphatic sentences with standard word order and intonation,[10] it holds for any degree of embedding that *the information of an embedded clause* has less communicative relevance than does the information of its matrix clause.

From thesis IV it follows that for all sentences of the specified type, depth of embedding is inversely proportional to communicative relevance.

As thesis I asserts, even simple sentences like (12a) express more than one piece of information; so this analysis should be continued into the syntax of simple sentences.[11]

DEFINITION iii. If two semantically equivalent sentences are broken down in the same way by means of the method of deletion, we say that they have *the same relevance distribution*. Sentences with the same relevance distribution we call *pragmatically equivalent*.

3. THE CONSTRUCTION OF COMPLEX INFORMATION IN DISCOURSE

Normally, sentences as complex as (7a) are used to summarize the information accumulated in the course of a longer conversation. Every complex sentence used in this way can be interpreted as a combination of the contents of consecutive utterances. From this point of view, the deletion method presented above appears as the converse of operations speakers engage in during the communication process. Instead of beginning with an already given complex of information and gradually breaking down its complexity, the interlocutors start with a simple utterance and collaborate in constructing more and more complex information. And, instead of deleting information as above, they add new information with each utterance to the content of the previous utterances. But the analogy between these procedures holds completely only when the newly added information is also of greater communicative relevance at the time of its utterance than the information that was uttered previously – just as, in the deletion method, the omitted piece of information is always of greater communicative relevance than the rest.

Whether this assumption is justified – and if so to what extent – can be discovered by examining a conversation among three persons A, B, and C, who utter the sentences (13) to (16) in turn.

(13)   (A:)   The soldiers crossed the border.

Sentence (13) provides complex information; it presupposes that "there are individuals that can be identified as soldiers" and asserts that "they have crossed a certain border". This claim is at the time of the utterance the main information and has the highest communicative relevance for the participants in the conversation.

Thereupon a new utterance follows:

(14)   (B:)   This surprised even the intelligence service.

Sentence (14) presupposes that "there is an organization that can be identified as an intelligence service" and describes the reaction of this organization to some given information. The information is identified in accordance with thesis III. It was neither surprising that "there were individuals" nor that "they were soldiers"; what was surprising was that "they crossed a certain border". The reference to this information can be made explicit through the parenthetical embedding of (13) in (14):

(14′)   This – i.e. that the soldiers crossed the border – surprised even the intelligence service.

In the present context sentence (14) is semantically and pragmatically equivalent to (14′). The information in sentence (13) has the function in (14′) of specifying the reference of the proform "this". As follows from thesis IV, all information in (13), including its main information, occurs in (14') as subsidiary information. The main information in (14') is that "this surprised even the intelligence service". Since (14') is semantically and pragmatically equivalent in the present context with (14), one may conclude that the main information in (13) is also implicitly used as subsidiary information in the utterance of (14). Thus, the step from the first to the second utterance in our conversation is tantamount to adding new main information and reducing the old main information to the level of subsidiary information with the function of specifying reference.

This analysis is confirmed by the role that all this information receives when a further utterance is added.

(15)   (C:)   That is serious.

In (15) the highest relevance belongs to the information that the state of affairs designated by "that" is regarded as "serious". Here again, the specification of reference for the proform "that" is accomplished in accordance with thesis III; the main information of the preceding sentence, "it surprised even the intelligence service", is directly commented on. If this commentatum is to be explicitly formulated, then the sentence representing it must be embedded in the comment sentence (15):

(15')   That – i.e. that this surprised even the intelligence service – is serious.

The embedding of (14) in (15') reduces the main information in (14) to the level of subsidiary information in (15'). This process of relevance reduction is repeated when we return to still older information in order to complete the specification of reference for "that":

(15'')   That – i.e. that this – i.e. that the soldiers crossed the border – surprised even the intelligence service – is serious.

By means of extraposition of the most deeply embedded clause we can formulate the information in (15'') in a way that is semantically and pragmatically equivalent and just as explicit but more easily understood:[12]

(15''')   That – i.e. that it surprised even the intelligence service that the soldiers crossed the border – is serious.

While (15') contains (14), (15'') contains the semantically and pragmatically equivalent sentence (14') in its place. The internal relevance distribution of (14') is not altered by embedding it in (15"). Evidently, relevance reduction has applied not only to its main information but also to its subsidiary information. Thus the main information in (13) is not only reduced in relevance once, when (13) is embedded in (14'), but a second time, when (14') is embedded in (15") – and this double relevance reduction can also be observed in the semantically and pragmatically equivalent sentences (15), (15'), and (15").

As a result of these processes we can state that in our conversation the degree of embedding of a piece of information reflects its time of utterance; to illustrate, the sequence of utterance of (13) and (14) is reflected by the degree of embedding of their respective information in (14'), and, similarly, the sequence of utterance of (13), (14), and (15) is reflected by the degree of embedding of their respective information in (15'). In each case, the matrix sentence is reserved for the information that is uttered last.

On the basis of thesis IV we know that in general, the less deeply information is embedded, the more relevant it is. So if the information uttered last in a conversation consisting of direct comments always occurs as the matrix sentence in a summary of this conversation, we can infer that the information uttered last is always of greater communicative relevance at the time of its utterance than the information of the previous utterances.

Let us analyse another direct comment in order to trace the way we came to this result. The text (13, 14, 15) can be continued by means of the utterance (16):

(16)   (B:)   That was also claimed by the commander.

In (16), the pro-sentence form "that" functions as an argument of the comment phrase "was also claimed by the commander". The reference of "that" in sentence (16) is specified by means of the main information of its preceding sentence (15); in parallel fashion, the reference of the pro-sentence form in (15) is specified by means of the main information of (14), the sentence preceeding it, etc. If the reference of a pro-sentence form is to be stated explicitly, then the corresponding information must be embedded to the right of it. Embedding reduces the communicative relevance of this information. The same thing also happens with its subsidiary information. It is on the basis of these operations, that the content of every newly added comment can assume the role of main information without changing the internal relevance distribution of the commentandum information.

Applying the previous procedures to summarize the course of the conversation between (13) and (16), we arrive at:

(16′)   That – i.e. that that – i.e. that this – i.e. that the soldiers crossed the border – surprised even the intelligence service – is serious – was also claimed by the commander.

If we extrapose as suggested for (15″), then we get:

(16″)   That it is serious that it surprised even the intelligence service that the soldiers crossed the border was also claimed by the commander.

(16″) is a possible paraphrase of (16), but (17), (18), and (19) are not, because they deviate from the structure of the previous conversation; they destroy both the semantic relationships among the pieces of information given by the predicate-argument relations in the comment sentences and the pragmatic relationships manifested in the sequence of the comment sentences:

(17)   That it was also claimed by the commander that it surprised even the intelligence service that the soldiers crossed the border is serious.

(18)   That it was also claimed by the commander that it is serious that the soldiers crossed the border surprised even the intelligence service.

(19)   That it is serious that the soldiers crossed the border was also claimed by the commander.

The object of the preceding analysis was a process of communication in which every following utterance directly comments on its preceding utterance. The results of the analysis can be synopsized in the following theses:

THESIS V. In a conversation consisting of a series of direct comments, the communicative relevance of every piece of information correlates with the *time of its first utterance*.

THESIS VI. In a conversation consisting of a series of *direct comments*, the relevance distribution of all information *remains constant*.

It is not at all trivial that the relevance distribution of the information uttered in a conversation should remain constant over the entire time covered; cf. (20, 21, 22):

(20)   (A:)   The soldiers have crossed the border, which surprised even the intelligence service.

(21)   (B:)   That this even surprised the intelligence service is serious, as was also claimed by the commander.

(22)   (C:)   That this was claimed by the commander is of no importance.

At the time of utterance of (20) the information "the soldiers crossed the border" is – according to thesis IV – communicatively more relevant than the information "this surprised even the intelligence service". However, if we specify the reference of the pro-sentence form "this" in (21) by embedding the corresponding information to the right, then we have:

(21′)   That this – i.e. that the soldiers crossed the border – surprised even the intelligence service is serious, as was also claimed by the commander.

or after extraposition:

(21″)   That it surprised even the intelligence service that the soldiers crossed the border is serious, as was also claimed by the commander.

And in (21″) the information "the soldiers crossed the border" is – again according to thesis IV – less relevant than the information "this surprised even the intelligence service": this also holds for (21) since (21) is semantically and pragmatically equivalent in the present context with (21′) and (21″). Thus we must conclude that the utterance of (21) has inverted the relevance distribution of the information in (20).

Something similar occurs in the transition from (21) to (22); according to thesis IV, the information that "that was also claimed by the commander" is less relevant at the time of (21)'s utterance than the information that "that is serious". If, however, we specify the reference of all pro-sentence forms in (22) by embedding the corresponding information to the right, then we get the following:

(22′)   That – i.e. that that – i.e. that this – i.e. that the soldiers

crossed the border – surprised even the intelligence service –
is serious – is also claimed by the commander – is of no
importance.

or after extraposition:

(22″)   That it is also claimed by the commander that it is serious
that it surprised even the intelligence service that the soldiers
crossed the border is of no importance.

If (22″) were a paraphrase of a sequence of direct comments, then these
comments would – according to thesis VI – have to be uttered in the
following order:

(13)   The soldiers crossed the border.
(14)   This surprised even the intelligence service.
(15)   That is serious.
(16)   That was also claimed by the commander.
(23)   That (however) is of no importance.

In fact, however, (22″) is semantically and pragmatically equivalent with
(22), and (22) is the last utterance of the conversation (20, 21, 22); so the
relevance distribution of the information must have been changed just
as much in the transition from (21) to (22) as in the transition from (20)
to (21).

Since the conversation (20, 21, 22) is a series of indirect comments, as
follows from Definition ii, the results of our analysis can be stated in the
following way:

THESIS VII. In a conversation consisting of a series of *indirect comments*, every piece of information added *changes the relevance distribution* within the complex information accumulated up to that point.

We have now advanced to a position where we can turn to the problem
of classifying types of dialogue.

4. GRAMMAR AND THE PSYCHOLOGY OF COMMUNICATION

A significant classification of types of dialogue must use criteria from the
psychology of communication. The example of comment texts will demonstrate how psychological concepts can be correlated with grammatical structures.

4.1. *Degree of Information and Competence of Judgment*

Let us imagine a conversation between two people who regard each other as equally well-informed and concede to each other equal capacities for sensible judgment. In standard situations, their dialogue will not proceed in such a way that one of them always presents new information and the other utters only comments on it. Rather, the number of comments and of utterances commented on will be quite similar for each of them.[13]

In a *report*, however, the dialogue will normally proceed quite differently. When only one of the interlocutors has to convey a large amount of news, he alone will supply the mass of utterances to be commented on whereas his partner will restrict his communicative actions to comments and questions.

DEFINITION iv. A conversation situation in which one participant provides the material for discussion and the other only comments on it is called a *one-sided dialogue*.

Besides reports, one-sided dialogues include, for example, university lectures of the old style, speeches before parliament, and litanies.

4.2. *Communicative Commitment*

If in a report the listener has only marginal interest in the information being conveyed, then the number of his comments will not exceed the minimum requirements of courtesy:

(24)   (A:) Yesterday the new managing head of the Social Democratic Party gave me an interview.

          (B:) That's interesting.

          (A:) He was on his way from the chancellor's office to the party headquarters. And I simply went up and spoke to him. I don't think he will have any more luck than his predecessor. It's difficult to commit such a big and so divided party to a single endeavor.

If, however, the addressee is committed to the topic, he will permanently interrupt the speaker to make contradicting or confirming, favorable or regretting, pleased or angry comments. Take the following dialogue:

(25)   (A:) Yesterday the new football coach gave me an interview.

(B:) Is that so!

(A:) He was on his way from the stadium to the training room.

(B:) You've got a good nose for hitting it just right!

(A:) And I simply went up and spoke to him.

(B:) That took courage.

(A:) I don't think he will have any more luck than his predecessor.

(B:) That's really news to me.

(A:) It's difficult to commit such a big and so divided team to a single endeavor.

(B:) That's probably true.

THESIS VIII. If the participants in a conversation consider each other equally well-informed and equally competent with regard to the matter under discussion, they measure their *communicative commitment* according to the frequency of their mutual comments.

4.3. *Intellectual Activity*

The intellectual activity of the addressee normally reveals itself in the degree to which he reproduces the formulations of his informant when commenting on them. Imagine the following interview situation:

(26)   (Journalist:) The Soviet Union and the Western Allies have guaranteed the free use of the transit highways to West Berlin in a joint agreement.

(People in the street:)

(A:) Bravo!

(B:) I had never expected that.

(C:) It is a source of relief to all Berliners that the Soviet Union and the Western Allies have guaranteed the free use of the transit highways to West Berlin in a joint agreement.

(D:) We are lucky that the Allies are here.

(E:) That an agreement was necessary in the first place is sad enough.

As is illustrated by the voices from the crowd the pieces of information supplied by the speaker can be repeated in the same formulation in which he uttered them (C) or they can be referred to by deictic (A) or by anaphoric (B) means; they can, however, also be rephrased. In this latter case the commentator can make explicit information that was given implicitly (D) or thematize information that was given concomitantly (E) in order to take a point of view about it.

THESIS IX. The *intellectual activity* of a participant in a conversation can be measured by how closely he conforms with the formulations of his informant in commenting on them.

4.4. *Integration of New Information*

For a person to come to grips with new information he must relate it to his previous knowledge and his previous attitudes. On the one hand, the hearer must give this new information a place in his individual system of knowledge; he can do this by establishing relationships with information he knows already; cf. (27):

(27)   (A:)   The Polish soccer team has qualified for the Olympic final.
       (B:)   They are even more successful than the newspapers had predicted.

On the other hand, the hearer must find a place for this new information in his individual system of values; he can do this by establishing relationships with interests he has; cf. (28):

(28)   (A:)   The Polish soccer team has qualified for the Olympic final.
       (B:)   That's great. So I shall see them play tomorrow. I have already bought a ticket for the final.

By expressing such relationships overtly in conversation one can influence the course of dialogue. An intellectually active and committed participant patterns the conversation according to his own process of reception by independently rephrasing parts of the information provided by his partner and confronting him with his own relevance distribution.

(29)   (A:)   Although I am generally, I think, a poor host, you will remember the party for years that I am going to give after the game.
       (B:)   You really don't believe that you are a poor host, do you?

By uttering an indirect comment B effects a redistribution of the points of emphasis in the complex information expressed by A. If A is not sufficiently detached, he will take up the topic introduced by B, as in (30):

(30)   (A:)   But the other day you said that I was a poor host.

Such an utterance shows that B's indirect comment has drawn A's attention away from his original topic of a football party. The discussion would have proceeded in an entirely different way if B had uttered a direct comment, as in (31):

(31)   (A:)   Although I am generally, I think, a poor host, you will remember the party for years that I am going to give after the game.
         (B:)   That's quite a promise.

This response leaves the initiative to A. He can continue the dialogue according to his own choice of topics. Note that the two courses of dialogue differ only in that B's comment is direct in one case and indirect in the other.

5. CLASSIFYING TYPES OF DIALOGUE

5.1. Active Dialogue

DEFINITION v. If a dialogue is not one-sided and both participants use only *indirect* comments, we call it an *active dialogue*.

An active dialogue allows all interlocutors great freedom in the selection of topics. It enables everybody to give the discussion a new turn without interrupting the train of thought.

The frequent redistribution of communicative relevance involved in an active dialogue is of great importance in education. A student able to evaluate independently the import of information given to him by his teacher can engage in strategies of active learning. If in addition, he is able to articulate his own preferences during the instruction by uttering indirect comments, he can determine which aspects of a problem should be given priority in the discussion and in which order and depth they should be dealt with. A class with more than one student conducting an active dialogue can effectively impede the teacher from inposing on the students his estimate of what is relevant in the information newly presented.

Practicing active dialogue can thus protect a child against an attitude of passive consumption long before he has attained the capacity of reflecting upon his own role in the learning process.

5.2. *Reactive Dialogue*

DEFINITION vi. If a dialogue is one-sided and the commentator utters only *direct* comments, we call it a *reactive dialogue*.

A person restricting his utterances to direct comments subjects himself unconditionally to the flow of information from outside. At each moment in the dialogue he regards as relevant whatever his informant makes appear relevant. His attention follows the path that his informant paves for him. This sort of passiveness can go so far that the commentator stops expecting a response to his or her comments; such comments take the form of mere reaction, and their content becomes negligible. And where the addressee does not articulate his process of reception sufficiently, the speaker can no longer assess the relevance that his individual statements have for the listener and adjust his strategy of speech accordingly.

(32)   (Knight-Templar)   [...] Not all are free who mock their chains.

        (Saladin)   A very apt remark! But truly Nathan, Nathan ...

        (Knight-Templar)   The worse superstition is to hold one's own for the most supportable ...

        (Saladin)   Perhaps! But Nathan ...

        (Knight-Templar)   ... to trust him ... and him alone ...

        (Saladin)   Good! But Nathan! – this is not Nathan's weakness [...]
                            (Lessing: Nathan the Wise, IV, iv)

In this dialogue, Saladin's only contribution consists in selecting a commenting operator. However, what can be commented on and what cannot is decided by the Knight-Templar by means of his formulations.

Reactive dialogues are frequently to be found in report scenes and master-servant conversations on the stage:

(33)   (Falstaff)   O, if I had had time to have made new liveries [...]. But 'tis no matter; this poor show doth better: this doth infer the zeal I had to see him, –

(Shallow)	It doth so.
(Falstaff)	It shows my earnestness of affection, –
(Shallow)	It doth so.
(Falstaff)	My devotion, –
(Shallow)	It doth, it doth, it doth.
(Falstaff)	As it were, to ride day and night, and not to deliberate, not to remember, not to have patience to shift me, –
(Shallow)	It is most certain.

(Shakespeare: Henry IV, part II, V. v)

As (33) demonstrates, even direct comments empty of content do have at least a phatic function; [14] they document to the speaker continuing interest and encourage him to continue. By changing emphasis they can even affect the course of the conversation; in our case, by exaggerating his agreement in the third comment ("It doth, it doth, it doth."), Shallow suggests to Falstaff that this topic be closed; Falstaff then reacts promptly by resuming the subject matter that gave rise to these remarks, his inappropriate clothing.

Because of the rigid distribution of commentandum and comment, the reactive dialogue is especially well suited for the establishment and exhibition of authority. The following example is instructive:

(34)   An Odd Bachelor Journeyman

Once upon a time there was a man who was pushing a wheelbarrow with a sack of rye. He was on the way to the mill to have the rye ground. He met another man there with whom he began a conversation.

– Good morning. Is this a good path?
– I have not yet given it a try.
– Is the mill going?
– It didn't pass by me.
– You are an odd bachelor journeyman.
– I am not a bachelor journeyman, I got married twenty years ago.
– Surely that was good.
– No, it was not so good. For the woman was terribly old.
– That was surely bad.

– No, it was not so bad. For she had a house and much money.

– Surely that was good.

– No, it was not so good. For most of the money was small change.

– That was surely bad.

– No, it was not so bad. For I bought four good pigs for these pennies.

– Surely that was good.

– No, it was not so good. For when our mother spilled the fat, the house burned to the ground.

– That was surely bad.

– No, it was not so bad. For I bought a new house instead.

– That was surely good.

– No, it was not so good. For when my old wife took a look at the house she fell down and broke her neck.

– That was surely bad.

– No, it was not so bad. For I took a new young wife.

– That was surely good.

– No, it was not so good. For she liked the young fellows more than me.

– That was surely bad.

– Yes, it certainly was bad. Good-bye.

(Svend Grundtvig/Denmark)

After the hair-splitting reaction of the "odd bachelor journeyman" to the questions that open the conversation, the other speaker limits himself exclusively to direct comments on his utterances. Since this pedantic fellow does not agree with even the simplest evaluation of his interlocutor, he allows himself to be provoked into telling his whole life's story in the form of corrections. Only a comment by the other speaker with which he agrees completely gives him the possibility to end the conversation with a farewell greeting.

5.3. *Direct Dialogue*

Direct comments only serve to fix relationships of authority if the distribution of commentandum and comment is one-sided. However, there is also a type of dialogue in which comments do not disturb the balance of role behavior between the participants in the conversation.

DEFINITION vii. If a dialogue is not one-sided and all participants use only direct comments in taking a point of view, we call it a *direct dialogue*.

In direct dialogues each interlocutor refers to the main information of the contribution of his predecessor. He integrates it into the context of his own utterance without redistribution of relevance. Direct comments are no longer understood as mere acclamations; rather their content is taken seriously and made the object of further direct comments: cf. (34):

(34)   (A:)   A member of parliament has been elected into our board of directors.
        (B:)   That is a great advantage for the company.
        (A:)   I don't think so.
        (B:)   That surprises me.

If the second speaker's utterance is comprised of more than one sentence, then the direct comment starts with the anaphoric representation or repetition of the last sentence.

(35)   (A:)   The subsidiary company didn't keep the promised date of delivery.
        (B:)   That is very disappointing. It ruins all our plans. These people know how much we are depending on them. If I had had my way, we would have produced the machine ourselves.
        (A:)   It's too late now for that. But we shouldn't give in yet. One of the competing suppliers has been trying for years to increase its share of the market. They will provide the machine soon.
        (B:)   That will not be easy for them. Their entire capacity is needed for their current production. But what can we do. Perhaps we should pay in advance.
        (A:)   I think we should.

In a direct dialogue, in contrast to the active and reactive dialogue, there is no collision of communicative interests; rather, everyone remains within the framework of a common train of thought and develops a common topic further. Therefore, the relative communicative relevance of the topics brought up for discussion cannot become an object of disagreement between the partners in the dialogue.

THESIS X. In an *active dialogue* no participant takes the communicative interest of the other into consideration, in a *reactive dialogue* only one does so, and in a *direct dialogue* both do.

THESIS XI. The *active commitment* of both participants in an active dialogue is opposed to the complementary play of *active and reactive commitment* in a reactive dialogue and the combination of the *reactive commitment* of both participants in a direct dialogue.

The following text by Bertolt Brecht, itself a direct dialogue, shows clearly, however, that even a direct dialogue neither presupposes *sameness* of interests nor guarantees *harmony* between the partners in a discussion:

> (33) Conversations
> "We can no longer speak to each other," Herr K. said to a man.
> "Why?" he asked with fright.
> "In your presence I can say nothing reasonable," complained Herr K.
> "But that doesn't matter to me," the other reassured him.
> "I believe you," Herr K. said bitterly, "but it matters to me."

Institut für Linguistik
Technische Universität Berlin

NOTES

[1] On the concept of communicative function cf. Posner 1979, introduction, p. 12 ff.

[2] Cf. Posner 1979, p. 166 ff; see also Posner 1978.

[3] An extended substantiation of this thesis can be found in Posner 1972a and in Posner 1979, pp. 28–55.

[4] All example sentences quoted in which no accent occurs are to be read with standard intonation.

[5] On the semiotic usage of the terms "syntactic," "semantic," and "pragmatic" see Morris 1938, cf. Posner 1972b. The concept of "pragmatic information in a sentence" is defined in Posner 1979, p. 12 ff.

[6] Where "relevance" is the subject of discussion in the following, this word always refers to communicative relevance.

[7] On the surface syntactic means of signaling communicative relevance cf. Posner 1979, p. 135 ff.

[8] On these terms cf. Curry 1963, p. 32, Bach 1968, p. 106, and Seuren 1968, p. 104 ff.

[9] Cf. Posner 1972a.

[10] The concepts of "standard word order" and "standard intonation" call attention to restrictions that have not yet been fully studied. In Posner 1979, p. 150 ff. the reader will find a more complete list of the syntactic parameters signaling what communicative relevance is assigned to the information expressed by the constituents of a given sentence. These parameters include:

1.   the position of the sentence accent
2.   the degree of embedding
3.   the syntactic category of the constituent in question
4.   word order
5.   the position and length of pauses in the intonation of the sentence

These five parameters themselves form a hierarchy according to their relevance.
[11] A method for this is described in Posner 1979, chapters 2 and 3.
[12] Cf. Langendoen 1967.
[13] Cf. Posner 1976.
[14] On the concept of the phatic function of speech acts cf. Malinowski 1923, p. 314 ff. and Jakobson 1960, p. 355.

REFERENCES

Bach, Emmon, 1968: 'Nouns and Noun Phrases', in *Universals in Linguistic Theory*. Ed. E. Bach and R. T. Harms. New York: Holt, Rinehart and Winston.

Curry, Haskell B., 1963: *Foundations of Mathematical Logic*. New York: McGraw-Hill.

Jakobson, Roman, 1960: 'Linguistics and Poetics', in *Style in Language*. Ed. Th. A. Sebeok. Cambridge/Mass.: U. P.

Langendoen, D. Terence, 1967: 'The Accessibility of Deep Structures', in *Readings in English Transformational Grammar*. Ed. R. A. Jacobs and P. S. Rosenbaum. Waltham/Mass., Toronto, and London: Ginn, 1970.

Malinowski, Bronislaw, 1923: 'The problem of meaning in primitive languages', in *The Meaning of Meaning*. Ed. C. K. Ogden and I. A. Richards. New York and London: Kegan Paul.

Morris, Charles W., 1938: *Foundations of the Theory of Signs*. London: The University of Chicago Press.

Posner, Roland, 1972a: 'Commenting: A Diagnostic Procedure for Semantic-pragmatic Sentence Representation', in *Poetics* 5 (1972), 67–88.

Posner, Roland, 1972b: 'Statt eines Vorworts', Introduction to the German translation of Charles W. Morris, 1938: *Grundlagen der Zeichentheorie/Zeichentheorie und Aesthetik*, Ed. F. Knilli. Munich: Hanser, 1972.

Posner, Roland, 1976: 'Discourse as a Means to Enlightenment – On the Theories of Rational Communication of Habermas and Albert', in *Language in Focus*. Ed. A. Kasher. Dordrecht/Holland: Reidel.

Posner, Roland, 1980: 'Semantics and Pragmatics of Sentence Connectives', in Speech Act Theory and Pragmatics. Ed. J. R. Searle, F. Kiefer, and M. Bierwisch. Dordrecht: Reidel.

Posner, Roland, 1979: *Theorie des Kommentierens – Eine Grundlagenstudie zur Semantik und Pragmatik*. Wiesbaden: Athenaion. First Edition Frankfurt: Athenäum, 1972.

Posner, Roland, 1980: *Rational Discourse and Poetic Communication – Methods of Linguistic, Literary and Philosophical Analysis*. The Hague and Paris: Mouton.
Seuren, Pieter A. M., 1969: *Operators and Nucleus – A Contribution to the Theory of Grammar*. Cambridge/England: U. P.

MARIAN PRZEŁĘCKI

CONCEPTUAL CONTINUITY
THROUGH THEORY CHANGES

I

The present paper, meant as a contribution to "theory of rational science development",[1] is concerned with the nature of conceptual changes characteristic of certain types of theory change. Some kind of conceptual continuity has usually been considered to be a necessary prerequisite of rational science development. That view, however, has been seriously questioned in recent years. The argumentation runs, roughly, as follows. Let T_1 and T_2 be two successive theories. Interpretation (sense and reference) of their specific terms is said to depend on the very theories in question. As the theories differ in what they say about their terms, the interpretation of the latter appears to be different within the two theories. In the case of mutually incompatible theories, their languages are claimed to be not intertranslatable, and their specific terms "incommensurable". But this seems to exclude the possibility of a logical comparison and rational choice between the two theories. To treat them as "rival", we have to regard them as speaking, partly at least, about the same subject-matter. And this seems to presuppose some kind of "commensurability" of their conceptual frameworks. The main aim of the following analysis is to supply some arguments in favour of a certain version of the "commensurability" thesis. Changes in interpretation of a theory's terms, occurring in the process of its development, may involve both changes in their sense and in their reference. What appears to be essential for comparing two successive theories, especially those couched in language of an extensional type, is some identity of reference. And this is just what I shall argue for in the present paper.[2] The argumentation adduced will be based on a specific conception of the semantics of empirical theories, which – from a formal point of view – may be called a model theoretic, and – from a philosophical viewpoint – an empiricist one. It coincides roughly with Carnap's semantics of empirical languages and theories.[3] It is thus subject to the well-known criticisms directed against that kind of approach. Its

R. Hilpinen (Ed.), Rationality in Science. 137–150.

justification and defence, however, must be undertaken at some other time and place.[4]

On the conception mentioned, any interpreted empirical language may be identified with a semantical system:

$$L = \langle F, A, M \rangle,$$

composed of a formal language F, and two classes of structures for F: $M \subseteq A$, A representing possible interpretations of L ("possible worlds"), M the actual interpretation of L (the "actual world"). A is thus conceived as a component determining the sense of L's expressions, M as a component determining their reference. According to a common practice, the sense of term t, $s(t)$, will be identified with the function which to any possible structure m for L assigns the denotation of term t in that structure, t^m:

$$s(t): \quad m \in A \to t^m.$$

The reference of term t, $r(t)$, will, in turn, be defined as the class of denotations of term t in all the actual structures for L:

$$r(t) = \{t^m\}_{m \in M}.$$

Our main concern here will be with the reference of terms, so understood. And the object of our analysis will be languages L_1 and L_2 of two empirical theories T_1 and T_2, which are taken to represent two successive stages in the development of what is usually called a given scientific theory. The interpretation of language L_i $(i = 1, 2)$ is assumed to be determined by the following assumptions, which constitute the main tenets of the empiricist standpoint adopted in the present paper.

(i) The non-logical vocabularies of languages L_i are assumed to contain a common sub-vocabulary of terms called non-theoretical with respect to both theories T_i – o-terms, for short. Their interpretation, being given in advance, is independent of any of these theories. It is assumed to be the same in both the theories. (The assumption will be somewhat weakened later on.) So L_1 and L_2 may be thought of as extensions of the same semantical system:

$$L_o = \langle F_o, A_o, M_o \rangle.$$

The assumption guarantees some kind of conceptual continuity between the two theories T_1 and T_2, viz. one restricted to their non-theoretical framework. Such an assumption seems plausible provided we conceive such theories in a sufficiently comprehensive way, incorporating within a given theory all sub-theories which underlie it (in particular, all the relevant theories of measurement). Descending deep enough, one may hope to reach a level which is neutral with respect to both theories and which may thus be assumed to be identical within both of them. When the above assumption is being questioned, this is usually due to a more restrictive conception of scientific theory, identifying it with the very top layer of the whole theoretical hierarchy involved.

(ii)   In addition to o-terms, the non-logical vocabulary of each language L_i is assumed to contain some terms called theoretical with respect to theory T_i – t-terms, for short. We shall here presume that they are symbolized by different symbols in different theories (say, as t_1 in T_1 and t_2 in T_2). The interpretation of t-terms in theory T_i is taken to depend on the very theory T_i. What this amounts to is, under the present approach, explicated as a condition identifying the meaning postulates for t-terms of language L_i with the so-called conventional component of theory T_i. Several definitions of that concept have been propounded in the recently published literature. Carnap's original proposal, applicable to all finitely axiomatizable theories, identifies the conventional component of theory T_i with the sentence $^R T_i \to T_i$, where $^R T_i$ represents the so-called Ramsey sentence for T_i (i.e., the existential closure of the formula obtained from T_i by simultaneous substitution of proper variables for all t-terms). Availing ourselves of the usual model theoretic notation (symbolizing, in particular, structures for L_o and L_i by m_o and m_i, the fragment of m_i corresponding to L_o by $m_i|_o$, and the class of models of a sentence, or set of sentences, X by $\text{Mod}(X)$), we shall then define the language L_i of theory T_i as a semantical system:

$$L_i = \langle F_i, A_i, M_i \rangle, \qquad \text{where}$$

$$A_i = \{m_i : m_i|_o \in A_o \quad \text{and} \quad m_i \in \text{Mod}(^R T_i \to T_i)\},$$

$$M_i = \{m_i : m_i|_o \in M_o \quad \text{and} \quad m_i \in \text{Mod}(^R T_i \to T_i)\}.$$

A_i so defined contains all expansions of structures in A_o to models of the conventional component of T_i; M_i – all expansions of structures in M_o to models of that component. The proposition sounds convincing provided there are no known reasons for supposing that the interpretation of t-terms in theory T_i depends, not on the whole theory T_i, but on some definite part of it, distinguished "from without", by certain pragmatic factors. No such reasons are assumed in the following.

II

The class M_i, which represents the actual interpretation of language L_i, may be shown to possess certain features which will prove decisive in our argumentation. What is meant here is the notorious multiplicity of any empirical interpretation. On the one hand, M_i is not to be identified with what is sometimes called the range of intended applications of theory T_i.[5] According to the traditional empiricist approach adhered to in this paper, M_i corresponds, not to the class of different particular applications, but rather to one maximal application, defined, e.g., as the union of all the intended applications. In spite of that, however, M_i cannot be identified with a single structure for language L_i, but with some class of such structures. This is caused by a feature which is characteristic of any empirical language: its fuzziness. It includes such phenomena as vagueness, in the case of qualitative concepts, and approximation, in the case of quantitative ones. Every qualitative empirical term, in particular, every empirical predicate, is, more or less, vague. This is a result of the way in which an empirical predicate acquires its interpretation. Any such procedure leaves some area of indeterminacy. For some objects there are provided no criteria of membership in the denotation of the predicate: it is not determined whether or not they belong to that denotation. So, besides positive and negative, there always exist some borderline cases of the given predicate. One way to account for this fact is to identify the interpretation of the predicate, not with a single set (or relation), but with some class of sets (or relations). Each of these sets (or relations) corresponds to a possible classification of all borderline cases of the given predicate into its positive and negative instances. This, in turn, results in identifying the actual interpretation of an empirical language, not with a single structure, but with some class of structures. Each of these structures contains, as the denotation of a given term, one member from the class constituting its interpretation.[6]

Languages of empirical theories, which are the object of our analysis, belong to quantitative languages. Besides qualitative terms such as empirical predicates, they typically contain some quantitative terms: symbols of empirical quantities. These, according to current theories of fundamental measurement, are introduced on the basis of certain qualitative empirical concepts. In consequence, vagueness of the latter affects the character of the former, being responsible for their intrinsically approximative nature. Since it is just the phenomenon of approximation that turns out to be decisive for our conclusions, let us examine it rather closely.[7] Theories of fundamental measurement define quantities as numerical representations of some empirical structures. In the case of extensive quantities, which constitute the main type of quantities referred to in empirical theories, those empirical structures belong to so-called extensive systems. Their essential component is an empirical ordering relation, characterized by standard axioms for extensive systems as a weak order in the universe of a given structure. (Besides an ordering relation, an extensive system is often conceived of as comprising an empirical concatenation operation, but this does not appear to be indispensable since its role may be played by the operation of set theoretic union.) Now, it is my contention that the ordering relation of an extensive system represents a typically vague concept, and that it is its vagueness that accounts for the approximative nature of the given quantity. The point, however, needs some comments since it has sometimes been misunderstood.

The ordering relation embodies some procedure of comparing two objects with regard to a given quantity by means of a certain measuring instrument. Every such instrument, and hence, every such procedure, are always imperfect in some respects. The kind of their imperfection that is usually raised in this context is said to consist in their limited sensitivity. No real measuring instrument can be perfectly sensitive. Every such instrument is insensitive to differences in the given respect smaller than some finite amount – the just noticeable difference, j.n.d. As a result of this, any ordering relation R_o defined directly by means of such measuring procedure lacks some formal properties required of ordering relations in extensive systems: in any sufficiently comprehensive set U there are some objects x, y, and z such that, though R_o does not hold between x and y and between y and z, it does hold between x and z. In effect, R_o is what is called a semiorder in U. xR_oy may be taken to mean that object x is noticeably greater in the given respect than object y.[8] That kind of imperfectness of any measuring procedure may, however, be got rid of by an ascent from the purely observational relation R_o to a more theoretical relation R, defined by means of the former as follows:

$$xRy \quad \text{iff} \quad \text{for every } z: \text{ if } zR_ox \text{ then } zR_oy, \quad \text{and}$$
$$\text{if } yR_oz \text{ then } xR_oz.$$

xRy may be interpreted as meaning that object x is at least as great in the given respect as object y. Defined as above, R can be shown to be a weak order in U, and thus to fulfil (together with a suitable concatenation operation) standard axioms for extensive systems. In this way we are able to overcome insensitivity of the given measuring procedure and inexactness of the corresponding measurement. If the set U contains objects with arbitrarily small differences in the given quantity, the accuracy with which we can measure the quantity is theoretically unlimited.

This seems, however, to contradict our actual scientific practice, which speaks for an approximative nature of any empirical quantity. To do justice to it, we have to realize that what is called imperfectness of a given measuring instrument, and measuring procedure, covers two different phenomena: insensitivity and vagueness. A measuring instrument may not react at all, or may react vaguely. It does not differentiate between certain objects, and differentiates vaguely between certain others. What does that vagueness consist of? First of all, there always seem to be, besides the determinate cases of the instrument's behavior, some indeterminate ones: cases where we cannot tell whether the instrument reacts, or not.[9] And if the procedure of comparing two objects is not confined to a single test, but includes a series of such tests (as it usually does in actual practice), there arises still another source of vagueness: when applied to certain objects, the instrument does not behave uniformly through all tests of the series – in some of them it reacts, in others it does not.[10] Now, all the cases of the instrument's "undecided" or "inconsistent" behavior, cases which yield indeterminate or dispersive results, make up the borderline cases of our original relation R_o. What kind of objects belong to these borderline cases? Whereas insensitivity of the given measuring procedure results in the existence of a finite j.n.d., vagueness of the procedure results in an inexactness with which that j.n.d. is being determined. It is to be thought of as equal, not to some real q, but to $q \pm \varepsilon$. And this amounts to the following: our procedure does differentiate between objects which differ in the given respect more than $q + \varepsilon$, does not differentiate between those which differ less than $q - \varepsilon$, and differentiates vaguely between those for which the difference ranges from $q - \varepsilon$ to $q + \varepsilon$. All pairs of objects of the third kind form the borderline cases of the relation R_o based on such procedure. Now, in contrast to insensitivity of a given procedure, its vagueness cannot be got rid of by transition from the relation R_o to the relation R, defined with its help in the way indicated. Vagueness of R_o is transmitted to R. It is not eliminated through that ascent, but only diminished partly. It is easy to see that only when the difference in the given quantity between x and y is greater than 2ε, the pair $\langle x, y \rangle$ presents a determinate, i.e. positive or negative, case of relation R. If that difference does not exceed 2ε, we are dealing with a borderline case of R.

On the approach to vagueness mentioned above, a vague relation is identified with some class of "sharp" relations, each of them corresponding to a possible classification of all the borderline cases into positive and negative instances. The counterparts of our vague relations R_o and R are thus some classes of "sharp" relations: R_o^* and R^*. Class R_o^* is here assumed to be restricted to those relations only which prove to be semiorders in the set U. Class R^*, which is to include all relations defined by the former in the way suggested above, will, in consequence, be composed of some weak orders in U. If we are now to treat all of them as our basic ordering relations, what we get, in effect, is a whole class of extensive systems S^*, each containing one relation from the class R^*. Class S^* appears, as a rule, to be even more comprehensive than class R^*, since the remaining components of an extensive system – especially an empirical concatenation operation (if such occurs, of course) – are hardly determined in a unique way, either. Now, to each extensive system S in S^* there corresponds one "exact" quantity, i.e. one real-valued function F, defined as a certain homomorphism of the given extensive system into a numerical structure of an appropriate type. Referring to standard extensive systems with concatenation operation o, this might be put more explicitly as follows:

If $S = \langle U, R, o \rangle$ is an extensive system and $u \in U$, then there is exactly one positive real-valued function F on U such that for any $x, y \in U$:

(i) xRy iff $F(x) \geqslant F(y)$:
(ii) $F(xoy) = F(x) + F(y)$;
(iii) $F(u) = 1$.

All functions defined in this way form a class F^*, which might be said to represent the approximate quantity measured by the given procedure. According to this approach then, an approximate quantity is nothing else but a certain class of exact quantities. The composition of class F^* is determined by the composition of class R^*. Taking into account what has been assumed of the latter, it is easy to realize what kind of functions will make up the class F^*. The values of two such functions for a given object will always lie within some fixed "interval of imprecision": for any function $F \in F^*$ and any object $x \in U$, $F(x) \in [k - \varepsilon, k + \varepsilon]$, for some k. This characteristic of class F^* seems to justify identifying it with what is usually called an approximate quantity.

Interpreting qualitative and quantitative empirical terms by some classes of their standard denotations – symbols of relations by classes of appropriate relations, symbols of functions by classes of functions of the proper type – results in identifying the actual interpretation of an empirical language L with some class of its structures, M. This, in particular, is true of our empirical language L_i and its sublanguage L_o. Both M_i and M_o are conceived of as classes of structures, containing more than one element. Such an approach to the concept of interpretation raises some problems concerning the notion of truth for languages so interpreted. There have been propounded some solutions to that problem, which we can only refer to in the present context.[11] But their main idea is quite simple. The true sentences in L_i are defined as sentences true in every structure in M_i. Sentences false in every such structure are said to be false in L_i. Those which are true in some structures in M_i and false in others are here qualified as neither true nor false in L_i. It is the concept of approximate truth that is meant to apply to this kind of statement. A sentence is thus said to be approximately true in L_i if it is true in some, at least, structure in M_i. A set of sentences, say T_i, is approximately true in L_i if all sentences of T_i are true in some structure in M_i. Let us see how this applies to a sentence which contains a function symbol f interpreted as referring to an approximate quantity in the sense here advanced, i.e. to a class of functions F^*. The sentence is true if it turns out true whichever function from those in F^* is chosen as f's interpretation; it is false if it is false under every such interpretation; otherwise, it is called approximately true. Thus, on the assumption that, for any $F \in F^*$, $F(a) \in [k - \varepsilon, k + \varepsilon]$, the sentence $f(a) \leqslant k + \varepsilon$ would be qualified as true, its negation $f(a) > k + \varepsilon$ as false, and the sentence $f(a) = k$ as approximately true. It is to be noticed that any exact quantitative statement of the latter kind may, in general, be only approximately true in L_i (unless it turns out to be a consequence of L_i's meaning postulates, as it is the case with the sentence $f(u) = 1$).

III

Let us now, under the assumptions adopted, compare the interpretations of two theoretical terms t_1 and t_2 in two theories T_1 and T_2, respectively. We shall concentrate on a case in which the problem of their "commensurability" becomes especially acute. This is the case of mutually incompatible theories. What is here meant by this may be explained as follows.

Theories T_1 and T_2 are incompatible if there are sentences V_o and W_o of language L_o such that $W_o \in Cn(T_1 \cup \{V_o\})$, $\neg W_o \in Cn(T_2 \cup \{V_o\})$, and V_o is true (i.e. true in all structures in M_o).

T_1 and T_2 are thus said to be incompatible if they entail contradictory o-statements on some additional o-assumption which is true (under its actual interpretation).

Now, it is easily seen that in such a case the senses of t_1 and t_2 can hardly be identical. If T_1 and T_2 are not logically equivalent, their conventional components ${}^R T_1 \to T_1$ and ${}^R T_2 \to T_2$ are neither so (except some trivial cases in which t-terms occur in the theories in an inessential way[12]), and, in consequence, $s(t_1) \neq s(t_2)$. This meaning variance, however, does not entail a corresponding reference variance. What we shall try to show is the fact that theoretical terms of two incompatible theories may well have identical (or partly identical) referents. The point, however, needs some explanation, since that identity may have quite trivial reasons. If ${}^R T_i$ is false in some structure in M_o, all expansions of that structure will belong to M_i, and the reference class of t_i, $r(t_i)$, will comprise all possible interpretations of that term (within the universe of M_o). This will, in effect, guarantee some kind of reference identity between t_1 and t_2: $r(t_1) = r(t_2)$, or at least $r(t_1) \cap r(t_2) \neq \varnothing$. But if a given structure in M_o is not a model of ${}^R T_i$, and hence is not expandable to a model of T_i, all its expansions to a structure for L_i, and so all interpretations of t_i, are – strictly speaking – unintended ones, as they do not satisfy the conditions expressed by T_i. So, when comparing $r(t_1)$ and $r(t_2)$, we should rather ignore them and restrict our attention to those interpretations of t_i which satisfy the conditions imposed by T_i. These will be called the intended interpretations in the strict sense, and defined as follows:

$$M_i^* = M_i \cap Mod(T_i),$$

$$r^*(t_i) = \{t_i^{m_i}\}_{m_i \in M_i^*}.$$

It is the intended reference classes $r^*(t_1)$ and $r^*(t_2)$ that are to be compared in our case.

Being incompatible, theories T_1 and T_2 cannot both be true (in the sense considered). And if one of them, say T_1, is false, there are no intended interpretations of its theoretical term t_1: $r^*(t_1) = \varnothing$, and the problem of reference comparison becomes trivial. The more so, if both of them turn out to be false. The multiple character of the classes M_o and M_i opens, however, still another possibility: both theories may be approximately true. As T_1 is incompatible with T_2, there is no structure in M_o in which both $^RT_1$ and $^RT_2$ would be true. But there may well be two different structures in M_o, m_o and n_o, such that $^RT_1$ is true in m_o and $^RT_2$ in n_o. The intended reference classes of t_1 and t_2 are then non-empty and the problem of their comparison remains open. Now, it is easy to point out cases in which $r^*(t_1)$ and $r^*(t_2)$ have some elements in common. Take the following schema:

$$T_1 = \bigwedge x((o_1(x) \to t_1(x)) \wedge (o_2(x) \to \neg t_1(x))),$$

$$T_2 = \bigwedge x((o_1(x) \to t_2(x)) \wedge (\neg o_2(x) \to \neg t_2(x))).$$

$^RT_1$ and $^RT_2$ reduce in this case to first-order sentences:

$$^RT_1 = \bigwedge x(o_1(x) \to \neg o_2(x)),$$

$$^RT_2 = \bigwedge x(o_1(x) \to o_2(x)).$$

Let M_o consist of structures m_o and n_o, defined as follows:

the universe of m_o = the universe of n_o = $\{a, b, c\}$;

$$o_1^{m_o} = o_1^{n_o} = \{b\}, \qquad o_2^{m_o} = \{c\}, \qquad o_2^{n_o} = \{b, c\}.$$

Obviously, $m_o \in \text{Mod}(^RT_1)$, $n_o \in \text{Mod}(^RT_2)$, and among their expansions to models of T_1 and T_2, respectively, there are structures $m_1 \in M_1^*$ and $m_2 \in M_2^*$ such that $t_1^{m_1} = t_2^{m_2} = \{b\}$. The theoretical terms t_1 and t_2 defined by two incompatible theories T_1 and T_2 are thus shown to have some intended referents in common: $r^*(t_1) \cap r^*(t_2) \neq \varnothing$.

The situation may, in general terms, be described as follows. Theories T_1 and T_2 characterize their t-terms t_1 and t_2 in two different ways. That difference, however, is compensated in structures such as m_1 and m_2 referred to above by suitable differences in interpreting some o-terms ($m_1|_o$ and $m_2|_o$ cannot thus coincide). This is possible if the differences required do not exceed the degree of indeterminacy with which the interpretation of o-terms has been fixed (both $m_1|_o$ and $m_2|_o$ must belong to M_o). It is into a similar schema that some type of so-called "correspondence" relationship seems to fit roughly.[13] The main difference seems to concern the common domain in which the two rival theories turn out to be approximately true. This domain has so far been identified with the whole universe of the two theories, i.e. the universe of the structures in M_o and M_i. Typically, however, it appears to be narrower. In contrast to the new theory T_2, the old theory T_1 usually proves to be false within its original universe. The domain in which the two theories turn out to be approximately true forms some subset only of the whole universe; let us denote it by U. What we shall thus compare are the intended interpretations of t_1 and t_2 within that subset only. To do this, we shall define a concept symbolized by $M_o(U)$ (or $M_i(U)$). $M_o(U)$ is the class of those substructures of the structures in M_o which have the universe U. So the interpretations of t_1 and t_2 being now compared are those belonging to structures in $M_1(U)^*$ and $M_2(U)^*$. As theories involved in a "correspondence" relationship are typically quantitative theories, let us suppose that o-terms and t-terms of theories T_1 and T_2 refer to some empirical quantities. To give a simple schematic example, we shall assume that in the intended structures in M_1^* and M_2^*, the terms t_1 and t_2 are defined by the following equations, which form part of the corresponding theories:

$$T_1: \quad t_1(x) = f_1(o_1(x), \ldots, o_n(x)),$$
$$T_2: \quad t_2(x) = f_2(o_1(x), \ldots, o_n(x)).$$

f_1 and f_2 are supposed to be different mathematical functions, that is, such that for some, at least, arguments $k_1, \ldots, k_n$: $f_1(k_1, \ldots, k_n) \neq f_2(k_1, \ldots, k_n)$. What is more, the above inequality is assumed to hold for all those arguments which represent the values of the o-functions in the actual structures for L_o. And so, for any $m_o \in M_o$ and any x from the universe of m_o: $f_1(o_1^{m_o}(x), \ldots, o_n^{m_o}(x)) \neq f_2(o_1^{m_o}(x), \ldots, o_n^{m_o}(x))$. Now, our suggestion is that there are $m_o, n_o \in M_o$ such that for any $x \in U$ (U being the domain in which T_1 and T_2 are approximately true): $f_1(o_1^{m_o}(x), \ldots, o_n^{m_o}(x)) = f_2(o_1^{n_o}(x), \ldots, o_n^{n_o}(x))$. This, of course, is possible only if for some, at least, o_i $(i = 1, \ldots, n)$: $o_i^{m_o}(x) \neq o_i^{n_o}(x)$. That difference in the interpretation of o_i makes up for the difference between f_1 and f_2. Our suggestion then implies that there are $m_1 \in M_1(U)^*$ and $m_2 \in M_2(U)^*$ (where $m_1|_o = m_0$, $m_2|_o = n_o$) such that: $t_1^{m_1} = t_2^{m_2}$. Some intended interpretations of t_1 and t_2 turn out to be identical within that subdomain in which both theories are approximately true. This fact is thus seen to be fairly consistent with the characteristics of the two theories given above.[14]

The case considered by us may be generalized so as to cover some other types of the "correspondence" relation. Let me here mention two such extensions only. One of them concerns the actual interpretation of the sublanguage L_o. It has been assumed to be identical in both the theories in question. But this seems unduly restrictive. There appear to be cases of theory development in which the interpretation of o-terms also undergoes some change: within the new theory T_2 they are interpreted in a more exact way than within the old theory T_1. We can account for this fact by splitting up the sublanguage L_o into two languages L_{o1} and L_{o2} with different classes of actual structures, viz.

$$M_{o2} \subset M_{o1},$$

and by defining the classes of actual structures for L_1 and L_2, M_1 and M_2, on the basis of M_{o1} and M_{o2}, respectively. Comparing the intended interpretations of t_1 and t_2, we shall here resort to those intended structures for L_1 only which are expansions of the structures in M_{o2}. When restricted in this way, they can be shown to stand in the same relations as before.

Another extension of the case considered thus far is meant to include such theory changes in which the new theory T_2 is said to be of "higher dimensionality" than the old theory T_1, containing terms with greater number of arguments than the corresponding terms of theory T_1. One can cope with such a situation by comparing, say, a k-place predicate t_1 of T_1, not with a corresponding $k + 1$-place predicate t_2 of T_2, but with a predicate t'_2 defined as follows:

$$t'_2(x_1, ..., x_k) \quad \text{iff} \quad \bigvee x_{k+1} t_2(x_1, ..., x_k, x_{k+1}).$$

Their comparison can then proceed as before.

These are examples only of modifications that might bring our analysis somewhat closer to situations characteristic of real theory changes. Our aim in analyzing some simplified schemas of such situations has been to show a possibility of sharing some common referents by theoretical terms defined by two rival theories. This seems to speak in favour of some kind of conceptual continuity in scientific development.

University of Warsaw

NOTES

[1] Such a theory is postulated as one of the objectives of the research programme "Foundations of Science and Ethics" in the Preamble to the programme.

[2] The main idea of the following argumentation has been presented in Przełęcki (1978).

[3] See, e.g., Carnap (1966). A model theoretic version of that conception, which is made use of in the present paper, has been developed, among others, in Przełęcki (1969) and Williams (1973).

[4] For some arguments on this point, see, e.g., Przełęcki (1969).

[5] See Sneed (1971).

[6] For a fuller presentation of that approach to vagueness and its semantics, see, e.g., Przełęcki (1969, 1976).

[7] A more detailed examination of the problems of approximate measurement is contained in Przełęcki (1979). For some account of theories of extensive systems, see, e.g., Suppes (1969).

[8] As an example of such a semiorder relation R_o one may take the relation of overbalancing, based on the procedure of weighing two objects by means of an equal-arm balance: object x overbalances object y iff when x and y are placed on the opposite pans of the balance, the pan with x descends relative to that with y. For an analysis of this example, see Przełęcki (1979).

[9] E.g., whether the pan of an equal-arm balance descends or not, in situations when it moves only slightly.

[10] E.g., in some tests in the given series of weighings the pan of the balance descends, in others it does not.

[11] See, e.g., Przełęcki (1969, 1976).

[12] See Przełęcki (1978).

[13] See, e.g., a characteristics of the "correspondence principle" in Krajewski (1977).

[14] The much discussed case of classical and relativistic mechanics seems to fall under a similar schema. To simplify matters greatly, the laws of motion characteristic of the two theories might be rendered as follows:

$$(1) \qquad f_1 = ma; \qquad (2) \quad f_2 = \frac{ma}{\sqrt{1 - \dfrac{v^2}{c^2}}}.$$

The classical law may, of course, be replaced by a law which contains v as an inessential argument, and, so modified, can be compared with the relativistic law according to the general schema given above. Our suggestion then is that, though the two laws differ, some intended interpretations of f_1 and f_2 may prove to be identical within some subdomain U (composed of objects moving with relatively low velocities), since for any object from U the value of v may be compensated by the difference in the values of m and a as determined by some intended interpretations of these terms.

REFERENCES

Carnap, R.: 1966, *Philosophical Foundations of Physics*, Basic Books, New York.

Krajewski, W.: 1977, *Correspondence Principle and Growth of Science*, Reidel, Dordrecht and Boston.

150 MARIAN PRZEŁĘCKI

Przełęcki, M.: 1969, *The Logic of Empirical Theories*, Routledge and Kegan Paul, London.
Przełęcki, M.: 1976, 'Fuzziness as multiplicity', *Erkenntnis* **10**, pp. 371–380.
Przełęcki, M.: 1978, 'Commensurable referents of incommensurable terms', *Acta Philosophica Fennica* **30**, (Issues 2–4: *The Logic and Epistemology of Scientific Change*), North-Holland, Amsterdam, pp. 347–365.
Przełęcki, M.: 1979, 'Some approaches to inexact measurement', *Poznan Studies in the Philosophy of the Sciences and the Humanities* (forthcoming).
Sneed, J. D.: 1971, *The Logical Structure of Mathematical Physics*, Reidel, Dordrecht.
Suppes, P.: 1969, *Studies in the Methodology and Foundations of Science*, Reidel, Dordrecht.
Williams, P. M.: 1973, 'On the logical relations between expressions of different theories', *The British Journal for the Philosophy of Science* **24**, pp. 357–367.

IVAN SUPEK

SCIENCE AND HUMANISM

SCIENCE AND ETHICS

Together with other social developments (such as the rationalization of the economics of money and commodities; and the resistance to mysticism and fanaticism), modern science has led to the absolutizing of objective research. This promoted, on the one hand, a view of the cosmos as independent of man, a cosmos with its own immanent characteristics and laws, and on the other hand, the empire of subjective feelings, opinions, and convictions. In the entire classical period, the scientific ideal was the discovery of the *true* laws or the ever more precise description of properties of the 'given' world. In this sense, research is reduced, in final analysis, to 'factography' – stating the facts, or, at least, presenting models which correspond to real processes. Hardly anyone suspected that this was not a self-evident approach, but rather only one particular human orientation.

This objective orientation strongly increased a non-partisanship in judging men and events, an aloofness from the tumult of everyday life, and, in general, that famous academic spirit which was a true haven for researchers in the era of wars of religion and conquest, of dogmatic pressures and aristocratic self-will. But at the same time it solidified an ethical indifference which has remained characteristic of scientists to this day. Ever since Galileo, following the advice of Cardinal Barberinni (who later, as Pope Urban VIII, became his jailer), declared that the science of mechanics did not affect church dogmas, positivism became a shield for the exact science – a very ineffective one, for the most part. The ruling circles were seldom satisfied with the indifference of thinkers, especially when science encroached on the field of the 'spiritual'.

This led to a sharp distinction between 'factographic' statements (which science would contain) and moral evaluations (which would be the domain of feeling or higher revelation). In this sense, the sentence 'The poppy is red' is characterized as factual, while the sentence 'The poppy is lovely' is taken to express the feeling of some individual. For certainly, it is claimed, everyone will observe directly (or with the help of

151

R. Hilipen (Ed.), Rationality in Science. 151–169.

a spectral device) that the poppy is red, while the flower does not have to appeal to a given individual. Similarly, this approach would reduce the question of whether someone would designate a given act as good or bad to a purely personal attitude. By confining itself to factographic or tautological expressions (and only these could be true or false), modern positivism has made a complete split between science and morality.

Can such a strict separation of the factual and and evaluative be maintained? It has already been noted that moral judgments cannot be made without knowing the real-life situation. Perhaps it is not too difficult to come to realize this. However, the belief in the purely factual nature or objectivity of scientific investigation is much more deeply entrenched. Using the earlier example of the poppy, we will first remark that the discrimination of colors did not come about for the sake of factographic description. Rather, in all likelihood, our ancestors would not have survived if they had not begun to distinguish colors. Therefore, color discrimination cannot be divorced from the rest of man's intentions and goals. Men did not develop thinking in order to be able to make a statement about this or that. They developed thinking first of all because they wanted to achieve some end or solve some life-problem (the satisfaction of hunger, defense against enemies, winning someone's favor, etc.). "This may all be so", a strict logician will reply, "but it still doesn't tell us anything about the meaning of the expressions themselves. Factuality becomes more conspicuous when one goes from the case of colors to the spatial properties of things." In confronting such general questions, we can let the history of science itself teach us.

Euclidean geometry and classical mechanics came into being after men had for thousands of years built all sorts of structures, measured land surfaces, made various kinds of objects, and used primitive machines. Geometric concepts and methods of reasoning cannot be separated from the use of rulers, compasses, and similar procedures for determining spatial relationships. In the same way, Newton's laws of force and of the relation between cause and effect have their roots in the entirety of human experience. The first scientific systems (geometry, mechanics, optics) were, indeed, discoveries, but discoveries most intimately linked with human existence, and without any doubt, their origins lie in the movement of our body, the work of our hands and eyes. The fact that the mutual interaction between men and nature can be reduced to fundamental and simple laws does not in itself mean these laws are purely objective or subjective. The fundamental unity of man and nature became still more evident when invisible fields and particles were discovered beyond the macroscopic reality. This atomic substructure makes itself known to us in its macroscopic effects and cannot be represented as a process in space and time. The investigator with his apparatuses cannot completely detach himself from the atomic substructure, so the laws of quantum theory apply precisely to this interaction between the investigator's instruments and the invisible particles. As the analyses by Bohr and Heisenberg have shown, it makes no sense to talk of the laws of some sort of things in themselves or of a pure object. If a basic unity is established at this level, then at the level of macroscopic physics man is just as indissolubly united to the entirety of nature.

Today we know the whole genealogy of science. From it we can see that individual disciplines are built on the basis of more fundamental systems (ultimately on geometry and mechanics) and that this entire developmental tree rests on certain invariant principles.[1] (See the diagram given below.) Newton's mechanics implied that momentum and

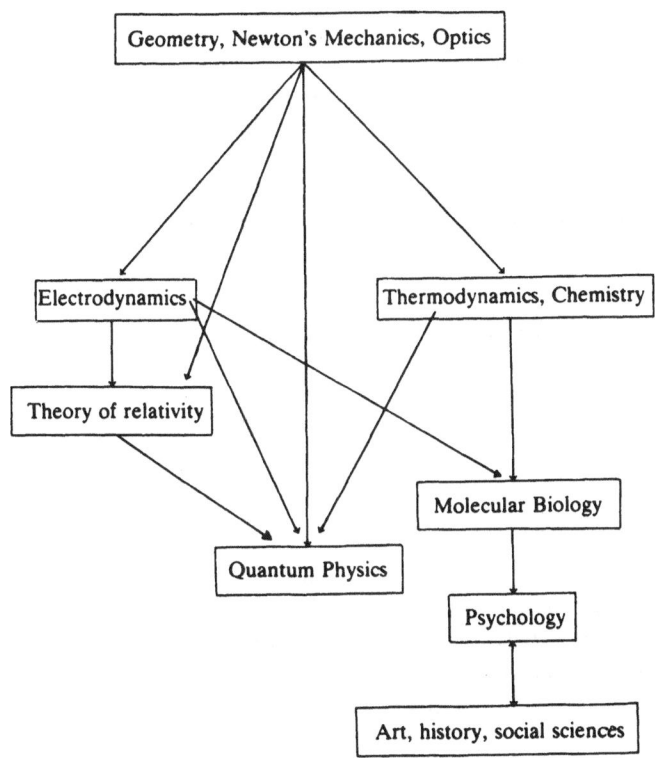

energy form a constant. When these principles were found not to hold in
the macroscopic realm, the principles were not abandoned, but rather,
action of invisible atoms and molecules was assumed. Science is not a
mere accumulation of facts, nor is a change in paradigms negotiated by
the experts. The entire genealogy of science refers to human existence
and activity in the world. In that existence and work one cannot sharply
separate the human body from nature, the objective from the subjective,
the theoretical from the empirical or actual.

Isn't the desire to know at the root of all philosophy, and doesn't it
guide the investigator? More than that, would men have survived at all
if they had not been curious? When we examine the human or any
other biological organism (and even mental functions), do we not find
so much of it arranged for some specific goal or purpose? How, then,
can we regard factography as the dominant characteristic of thought?

The introduction of *force* in physics has been accompanied by violent
theological and ethical controversies. Galileo in his famous *Dialogo*
attacks Marcus Antonius de Dominis because the latter explained tides
by attraction force from the moon. Newton's inclination towards theology and even alchemy (together with Kepler's rules about the movements of planets) was decisive for imagining universal forces between
bodies and atoms. Without such deep insights and evaluations no
scientifical system has been erected. What once was an ethical challenge
can later appear completely indifferent. The ethical content of scientific
innovations is related to the general view of the world and to social
circumstances.

INDUCTIVE GENERALIZATION, HYPOTHESIS AND THEORY; THE PROBLEM OF TRUTH

Much of scientific research is based on generalizations such as "all
ravens are black"; put in logical form, if an individual x has the
property F, it also has the property G:

$$(x)(Fx \rightarrow Gx).$$

Putting inductive generalizations and physical laws in the same logical form is the beginning of all confusions. First of all, physical laws
express relations between physical quantities which are measured and
formulated in accordance with some invariance principles. Physical law
implies always a mathematical form and concepts about the cosmos. So

much for the difference between simple inductive generalizations and physical laws. In scientific research we clearly distinguish inductive generalizations, hypotheses, and theories. The best example of an induction is Ritz's rule that frequencies of all spectral lines are given by the differences of characteristic terms;

$$v = T_n - T_m.$$

For any element and compound the characteristic terms have been established with fine precision. If a philosopher (like K. Popper) would deny the importance of induction, he would make a fool of himself among researchers. But Ritz's rule says nothing about atoms. It was Bohr who first saw in the characteristic terms the discrete energies of atomic systems:

$$E_n = -hT_n \qquad (h \text{ is Planck's constant})$$

This hypothesis of Bohr's led Heisenberg to the discovery of matrix mechanics, which is a full-fledged theory, with a system of equations which enabled us to calculate the micro-processes.

Mendel's well-known rules originate also from accurate observations like Ritz's rule. Again, the explanation of these rules by the existence of genes in the living cell is a hypothesis. The full theory is established later by molecular biology.

A scientific hypothesis is usually connected with an inductive generalization, but it leads to a deeper structure and so it becomes a constructive element of a theory.

The three levels (inductive generalization, hypothesis, and theory) are not always as well distinguishable as they are in physics. It would be wrong to regard induction as a pure jump from the factual.[2] Inductive generalizations also involve some theoretical concepts. The transition is really from a lower branch to a higher branch of science in the genealogy of science.

Going from classical physics to Quantum Theory, we have to change the basic concepts of physical law, causality, and subject-object relation. According to Hume and even Kant, causality could be expressed in the form: An event B follows necessarily from an event A, and C follows necessarily from B, etc. But as Hume stated, we can never see from A why the event B has to follow. The determinism or causal connection in physics is of a different kind. We have some functional relations between physical quantities, such as Newton's law, Maxwell equations, two principles of thermodynamics, etc. The solar system gives us an example where we could substitute real objects (sun and the planets) for the terms of an equation between physical quantities, believing that other properties of the objects are not relevant. This created often confusion for a realist who then believed that the Newton equation is a relation between objects, and similar to an inductive generalization. If we would like to treat more complex bodies and their interaction, we have to take into account additional physical laws (mechanics, electrodynamics, thermodynamics):

$$f_1(p, q, m \ldots) = 0$$
$$f_2(E, H, c \ldots) = 0$$
. . . .

If the relevant properties of the objects could be identified with the physical quantities in the equations above, and no other properties have to be taken into account, then we could have a deterministic system. Due to the margin of error or various unknown properties (even atomic structure), no deterministic case could be actually established, except in the functioning and description of the apparatuses which have been built according to definite physical concepts and laws for well-defined measurement. Sometimes such an apparatus, e.g. a GM-counter, could only give a "yes" or "no" reply to physicist. It has often been said that the understanding and measurement of microprocesses presupposes the determinism of macrosystems. This may be a too strong a requirement. In our weaker formulation determinism is required only for physical apparatuses (allowing 'margin of error' quite different from quantum possibilities, e.g., two possible orientations of spin in magnetic field).

PROBLEM OF TRUTH

The basic problem of the theory of knowledge has often been seen as the relation between our concepts (thoughts) and outside reality. This relation is for Aristotle the criterion of truth:

to say of what is that it is	= true
to say of what is that it is not	= false
to say of what is not that it is	= false
to say of what is not that it is not	= true

In accordance with such a realism Ludwig Boltzmann regarded the development of physics as a more and more accurate picturing of outside things. "I hold the view that the task of theory is the construction of a picture of the external world ... It is a peculiar tendency of the human mind to create such a picture and to fit it more and more to the external world ... The continuous perfection of the picture is the main task of the theory." (The lecture 'On the Importance of Theories', 1890, translated by E. Broda, *International Boltzmann Symposium*, Vienna 1972). Similarly, B. Russell initially looked for a real thing for any logical subject, and the young Wittgenstein in the *Tractatus* explained elementary sentences as pictures of states of affairs. The original difficulty of this *mirror-* or *correspondence-theory* lies in the concept of an *outside* thing. The *outside* of a box is a reasonable concept. Exercising some care, one can even speak about bodies outside or around one's own body. However, to speak of *the physical reality outside of our mind* is misleading. Physical concepts and laws as we can understand from the genealogy of science, are rooted in human actions and are related to the interactions between men (society) and nature (environment).

When truth is degraded to mere trivial statements of observations, then it is possible to lose its connection to ethics. However, the search for the truth which resolves, unifies, and advances the whole complex of problems and undertakings is the motive of the investigation.

If, while in a café, I say: "There are no witches", my friends would not be likely to pay much attention to such a statement. Nobody would regard this statement as a matter of truth or morality. But suppose I were in a 16th century market place where a woman was being brought to be burnt at the stake. If I were then to shout: 'There are no witches!', people would understand that I was protesting the burning of an innocent woman and saying 'You must not do that!' This example shows the impossibility of a strict separation of the so-called factual statements from the imperatives.

Our perplexity in distinguishing between factual statements and moral judgments would be even greater if we were to go from our familiar contemporary language to the scripture of the ancient Egyptians. What we would regard as myth or poetry the ancient peoples would consider to be an adequate description of some event.

It is not possible to draw a strict dividing line between accurate thinking and imagination. Imagination plays the main role in science as well as in the arts, A scientist does not only collect facts (as a botanist mainly does), but selects, supplements, orders, and explains various observations or experiments. Moreover these observations and experiments are connected with his concepts and previous theories. In scientific research it is impossible to separate pure empirical data from our conceptual framework. When I say "Peter is in the garden", the concept of an inside and outside, as of a box, is implied. Even the most trivial of our statements presuppose such a conceptual framework. This framework is certainly rooted in human existence, in the interaction between men and nature.

From this point of view a sentence such as "The chalk is white" is not as simple and independent as Russell thought. The ability to differentiate colors was developed during the long hunting period in human evolution when man's curiosity about his environment and mutual understanding were essential for the survival of the species. To understand what 'white' is we have to know red, yellow, blue, etc. We could not distinguish a piece of chalk from anything else without having made and used such things. To give our statement an immanent characteristic of *truth or falsity* outside a given context is an unallowable generalization, even though, in many cases, it is practically valid in our daily speech. If I say "Peter is curious", I am describing his behavior, a fact. But Peter is anxious to see circumstances in a camp of political prisoners. If I then say "Peter is curious", I am at the same time expressing my moral admiration or disapproval. Even such a factual word as 'white' often designates a moral value. The dividing line between factual statements and moral evaluations cannot be maintained. Similarly many factual statements are covering imperatives. For instance, when a teacher in a polluted classroom exclaims: "Windows are closed", pupils will immediately conceive the command "Open the windows!".

Characterizing every statement as true or false implies a naive or a Platonic ontology. According to standard theories of knowledge a true statement is related to an existing state of affairs, and a false statement to a non-existing state of affairs. Such a conception of truth presupposes an autonomous reality with definable properties and relations which we grasp more and more accurately in our research.[3] Classical physics partially presupposed such a naive ontology, but quantum theory has abandoned it.[4] In quantum theory all properties and relations are rooted in the interaction between men and nature. For this reason we cannot use an ontological criterion which is independent of human inquiry or existence. Truth is embodied in the dense tissue of our procedures and thoughts. 'Truth' is that which is suitable, that which fits the whole field of investigation, that which expands our concept of reality.

This interpretation of truth is intimately linked to everyday experience. Men consider those instructions or ideas to be true which lead them to their desired goals or which aid their successful actions. This is not a case of praxis confirming thought, but rather the point is there would be no praxis at all without thought, just as there would be no thought without human action. Truth is mated to action; its consequences can be tested by external, public effects; it is the reliable guide and résumé of our actions.

According to the correlation hypothesis, the succession of physical states in brain would be accompanied by succession of mental states. Because the physical states are causally connected the mental states cannot be more than a shadow of matter. Ludwig Boltzmann elaborated the famous *mirror theory*; the consciousness is a reflexion of the movement of the matter; by Darwin's selection human race acquired *right* pictures about environment and the so called "laws of thought".[5] Caught in the trap of classical physics, without understanding of quantum theory, Sayre comes to the conclusion:[6] "The identity theorists maintain, as we come to understand brain processes even better we will find that consciousness is nothing more than a type of neural occurrence." Typical mistakes could be also found in various articles by Eccles, especially in his *Brain and Consciousness Experience*. The physicalism of Quine and Smart shows the same ignorance of Quantum Theory. The simple reduction of biological and mental processes to physical processes is a gross misunderstanding of the range of physics. Physics starts with space-time-force relations and consequently all the measurements are expressed in meter-second-gram units. Quantum Theory is primarily concerned with the transformation of the invisible reality into our familiar macro-framework. According to Heisenberg's indeterminacy principle we are no more able to represent underlying atomic processes as taking place in space and time. With the disappearance of "pure object" disappears also "the body". The psycho-physical parallelism and the identity theory have lost their meaning; at bottom, neither physical nor mental states are determinable. The application of physics to biology and psychology meets also the difficulty that we cannot separate in a living organism (in a strict sense) macro-structure from micro-processes as we do in physical experiments. For this reason the simple physical concepts and statistics cannot without reservation be expanded to biology. This complication has been missed by J. Monod[7] in his statistical explanation of evolution.

At its height, truth creates a dramatic tension and moral appeal. The discoveries of Copernicus, Galileo, Darwin, or Heisenberg did not mean only a change of theory. They came into conflict with the dominant views and caused many moral dilemmas. We admire Bruno for dying for the new truth, but are unhappy about the weakness of Galileo. Scientists have to stand up and be counted when the truth of their research is at issue. Science is the extension of primitive human curiosity. Is "curiosity" a moral term? Of course, the curiosity of our neighbor can be a nuisance to us. But on the other hand, would we regard a man as moral if he would be completely disinterested in what is happening around him and unconcerned about the misery of the world? Moral behavior is deeply rooted in everyday life and human social existence. Morality is related to human beings and their deeds, but not to a man's acts or their consequences taken separately or individually. The goal of morality is a *good human being*, one who has compassion, love, and understanding for both those nearest to him and those farthest from him.

In the light of all we have said, it would be hard to accept the position that science is ethically neutral and that morality is independent of facts. The truth is an expansion of our reality and includes all the moral involvements of that reality. A good man has to be a truthful person. Ultimately, truth has an ethical appeal.

ART AND REALITY

While scientific investigation strives for general structures and laws artistic creation centers on persons. Stimulated by situations and conflicts, the artist resonates to the more delicate and complex. Many things which are merely hinted at or in embryo are brought to full openess by the artistic work. It is this all-embracing conception which, for the most part, constitutes the art work. In the artwork, everyday elements and components from the play of the imagination are reworked and combined into a totally new whole. Creative artworks are radically different from natural objects and social products. While the planets are studied in their mutual movements, and the production of trucks depends on a specific relation to the state and needs of industry as a whole, Raphael's 'Madonna' has a meaning only in relation to the painter who made it and those men who take delight in it. Of course, it is undeniable that such a painting was made through a knowledge of the developments in the art of painting, but the painting itself, as it hangs on the wall, is not at all dependent on the presence of other artworks, and its effect is produced directly on each man. A locomotive runs on tracks, pulls passenger cars, and stops at scheduled stations. Similar interdependencies are observable among other objects of material production. By contrast, an artwork does not serve any material function, and does not stand in any direct relationship to other objects. For ages, works of art have been made in order to transmit the artist's thoughts, visions and feelings to the audience. If they fail in this, they are lifeless.

Art enormously expands human reality. If we made the experiment of wiping out everything dealing with various forms of decoration, ritual ceremonies, sculpture, dancing, music, theater, and poetry, hardly anything would be left of our everyday language. Its true, in Plato's judgment, that art imitates reality, but it by no means stops with that. In addition to what is real, as Aristotle points out, it also creates that which merely could be (and we would add: also that which cannot actually be). If a painter or poet didn't use elements of existing reality, he would not succeed in making contact with his public. However, these real elements need not be unchanged, nor need they be combined into a whole in a real-life way. Human imagination can augment or diminish anything; it can transpose or combine things; it can modify or transform them; project them into completely different environments; join disparate parts or decompose integral units. In short, this playing, sometimes as free as a dream, sometimes strictly mathematical, carried on by the artist's conception or conflict, results in an artwork which evokes similar mental processes in the viewer, reader, or performer. Paintings, sculptures, dances, symphonies, dramas, and novels create an unique sublime, but very intensive, life which fundamentally transforms and exalts all human thinking and feeling. This 'feedback action' is already so deeply ingrained in daily life itself that it can hardly be separated from the directly practical and real. The esthetic sense is just as deeply rooted in the primordial community as is the ethical sense.

Open as it is to everything that is possible and impossible, art is the most radical challenge to a petrified society. Power-holders have always been concerned to prevent their subjects from even imagining a different life, much less attempting to change the world. It is true, as Schopenhauer said of music, that when one becomes resigned to the impossibility of any improvement, esthetic enjoyment becomes a comfort to man. But such a substitution for real life cannot for long satisfy everyone. Even in his day, Byron felt the fire in poetry which drives an author to be a torch-bearer in the darkness. How is one to sing of beauty, purity,

and freedom when everything around one is ugly, base, and enslaved?
The best plays of Sophocles and Ibsen are saturated with a deep moral
appeal, while Shakespeare, even when he does not moralize, is a peerless
humanist to whom 'nothing human is alien', to use a motto of his time.
Ruling regimes have always tried anew to reduce men to performers of
set tasks, and by now, we have indeed come a long way in such imper-
sonalization, specialization, and mass production. Literature, by con-
ceiving and bringing to life all sorts of conditions and people, makes
every man conscious of his manifold possibilities and prepares him to
revolt against the existing order. Even though painters, composers, and
poets seldom offer detailed instructions for action, they still prepare a
favorable climate in which ideas of reform flourish, even ideas of a
complete rejection of existing reality. It is, therefore, no wonder that
every dictatorship has placed art under control – with total impoverish-
ment as the result. Art flourishes through opposition to tyranny, oppres-
sion, cliches, brutality, banality, alienation, and depersonalization. For
the arts, freedom is what air is to the lungs. Nowhere is tradition, in the
best sense of the term, as strong as in the arts. Nowhere else do men
dream up new worlds as vividly. Painters and writers dig up the most
hidden roots in the soil of their history; they penetrate the furthest
hiding places of the soul. And while reviving the legacy of the past, their
conjectures give leaves to the seed of a fantastic spring. This conflict of
the traditional and the new moves the artist now backward, now
forward. It is no wonder, then, to find so many conservatives and rebels
among artists. In the greatest of them, these two forces are united in a
complex vision. Who would ever feel truly rebellious if he did not also
love what surrounds him?

The arts offer us an insight into the most hidden human motivations.
The artist has the ability to think-with and feel-with men in his
imagined encounters, in the adventures of his imagination. This imagin-
ativeness is an essential characteristic of human thought and is an
organic extension of the accurate or factographic description. We could
hardly understand any individual statement of another man if his
accompanying and encompassing thoughts did not resonate in us. There
is no accurate thinking without imagination, and the boundary of the
seeable is merely the point at which events can be defined and causally
linked in space and time. Artistic truth is, in the first place, the authentic
expression of the psychological flow, freed from automatic conventions
and banal cliches; mostly it is the flow which reaches down to the

deepest contacts of man and the world. While factographics writes down events as they play themselves out in the space-time arena, and so is aimed at the past and the present, the artist expresses his own conceptions with their full emotive extension. External locations and encounters generally enter only as re-worked memories. To an extent, popular wisdom confirms this when it says "Art stands outside time", or in another formulation, "Art is eternal". In this regard we must always keep in mind that the artist himself, as well as his motivations, is influenced by his times, and that the creative activity itself lasts since old times.

Even though it is imagination which raises us to the plane of art, if it is basically corrupted it degenerates to lies, willingly or unwillingly. Special interest, advantages, prejudices, and fetishes can all cause some fantasy to be so lodged in real spatial-temporal circumstances as to seem to be the truth. Such kinds of lies succumb to investigation, and a more careful consideration of the spatial-temporal web refutes them. For this reason, those who get involved with this type of lie use all means to prevent subjecting their allegations to empirical investigation.

Balzac once compared himself with an eye which looks at novels written by the French society. The aesthetics of socialist realism was based on *mirror theory*: Consciousness, and particularly artistic vision, is a reflexion of the movement of matter with all its dialectical conflicts. Restricting or even denying the freedom of creativity the ideology very soon came into conflict with science and art.[8] When such ideological dogmas were combined with political pressure, the results were regrettable.

The question of truth is much more portentious today than it was in civilizations which changed slowly, in which customs, work habits, and social structures lasted for centuries. In such eras people believed that laws determined the course and state of the world. In our day, however, newspapers, films, and television have increased the spiritual unrest at a time when art itself had already unsettled the human spirit and the changes in science had raised human work to a completely new plane, resulting in a total threat to all life. In such total jeopardy and change, one loses hope in any certitude or truth, and such a loss immediately plunges one into a new groping.

Anyone who seeks to find in truth a mirror of things, a determined law of the cosmos, or even an eternal essence of that which exists, will be quickly overwhelmed by authentic human investigation and creativity.

There is no truth in a predestined world. There isn't simply because there is no such world, nor would such a world have the least meaning for human knowledge.

Without truth and freedom man would be lost in the world. They are characteristic of thought and feeling, of the entire spectrum of diverse desires to know, constructive abilities, imagination, adventuresomeness, and contemplation which goes on in our leisure. We feel every diminution of truth and freedom as the greatest and deepest injury. This is why we welcome a world in which the fundamental human desires, highly developed by sciences and arts, flourish ever more strongly and sophisticatedly. This truth is indissolubly linked to human world and experimental procedures. When a specific thought is so incorporated into the totality of actions that we can at will repeat, begin, continue, and finish it with the same results, then we consider that thought clearly defined and true. The fact that there are such general laws is the discovery which grounds our theory of knowledge.

The arts, like the sciences, are the highest arches of man's love of knowledge, his imagination, and his ability to construct. Through them we become more completely aware of all that man can be, do, and know. What geniuses like Bach, Mozart, and Beethoven achieved in music; Michelangelo, Cezanne, and Picasso in painting; Dante, Shakespeare, and Dostoievski in literature; Newton, Darwin, and Einstein in the natural sciences can hardly be surpassed in the future. Since it starts with human creativity, humanism cannot take an ultimately negative stance towards the past. If it did so, it would lose its understanding of man. Humanists adopt neither nihilism nor utopianism, both of which too easily discard all that has thus far been created, and so the very bridges to a better world. What human imagination, sympathy, and creativity have given us thus far is sufficiently great to serve as an anchor of our hope.

Artistic and scientific creativity differs fundamentally from material production which constantly increases (and destroys) all kinds of commodities, means of production and commerce, and the networks of mass media and control. While such material production is coming in conflict with the energy and raw material potential of our planet, there are no limits to the arts and sciences. Society may have to restrict or even completely ban overly-ambitious or perverted applications of genuine scientific research, but man's love of knowledge, imagination, play-

fulness, and fancy itself will not thereby be impaired. On the contrary!
We can hope that precisely the increase in such creativity and play-
fulness will help produce an historic change of course for our entire
civilization – from feverish consumption to a more natural, deeper, and
spiritual enjoyment. Only this will avert the disaster to which our hyper-
industrialized world is rushing headlong.

PHILOSOPHICAL APPEAL

The universality of moral appeals is particularly critical in a time when
applications of science, under the tutorship of political-military leaders,
have brought the entire *Earth* to the brink of total catastrophe. When in
June 1944 at the Congress of Partizan Intellectuals in liberated Croatia
we anticipated the imminent use of the atomic bomb with its potential
for the complete destruction of mankind, and thereupon immediately
concluded that co-existence and disarmament were necessary, others of
those present dismissed the danger out of hand as completely unlikely,
and considered our conclusion anti-revolutionary, contrary to revolutionary dialectic of marxism. Fourteen months later, when the first
bombs were dropped and when the development of the hydrogen bomb
made my anticipation a frightening reality, the triumph of revolution
and progressive forces, at least by means of armaments, appeared ever
less of a historical law. Once both nuclear powers have at their disposal
the potential of total distruction, how can one dare to still prate of the
necessity of war, or of the victory of one or the other of the two sides?
At the end of the Second World War we were confronted with the task
of transforming, or cutting off at their very root the then current political conceptions or ideologies while still preserving all the fundamental
desires which filled the hearts of humanists, socialists, and other strivers
for social justice. Clearly no peace-making appeal would be successful if
it resulted in preserving the contemporary world with its overflowing
inequalities, exploitations, and violence.

Since the principle of general complete disarmament of a united world
had become a question of 'being or not being' for all men, both today
and tomorrow, and since we had already spent more than three decades
getting that knowledge accepted by even otherwise militant citizens, our
first concern became how to overcome the obstacles and reach that
undebatably desirable goal. Our endeavor is so much the more urgent

because the politics of force results only in destruction and suicidal crises. The moment is ripe for mankind's reason and fellow-feeling to fulfill their mission of salvation.

However, this must not remain an abstract, inefficacious desire. It is only when millions of steps begin to be taken toward the given goal, and each of those steps is specific and concrete, that we will make progress. It would be senseless to expect that everyone would have the same viewpoints in regard to the day-by-day conduct of politics or the organization of daily affairs. The important thing here is that no one outlaw the thoughts of others; that no minority impose its program and social system by violence on a majority. Culture, which rests fundamentally on the freedom of creativity and acceptance, must today more than ever be opposed to every form of force or violence whether in the relations between states or individual men. (This, of course, does not mean a permissiveness toward criminality which is a violent act and an injury of other persons.) When political parties and religious groups have accepted this principle, we will certainly be closer to solving the contemporary crisis. Living on an atomic volcano, a single fanatical spark could call forth a catastrophic explosion. Tolerance is not only an imperative of investigation, but is the condition of general existence. But that tolerance does not mean indifference; if it did, it would produce little benefit. Longing today for a disarmed and united world, we must necessarily accept the diversity and self-identity of its individual parts. If we would attempt to stitch together into a single program the east and the west, the north and the south, in all likelihood we could expect only a conflagration.

Warnings of the danger of total destruction were insufficient to induce states to adopt general and total disarmament even though they declared this their goal at the United Nations. It is clear that military blocks, and even the lesser powers, will not give up armaments as long as their positions of power remained based on armed might. Secrecy and police control in the production of atomic weapons have strengthened the most oppressive forces in any society; our country has not been spared from this trend when the head of State Police became the President of the Atomic Commission. This was the greatest challenge to our humanism. Unfortunately too many intellectuals remained silent or even sympathetic to such a conspiracy, mainly for nationalistic feelings and party ambitions. It is essential to create such frames of mind and relations between nations as would exist in a disarmed and united world.

Without a deep, democratic transformation in every land, 'foreign politics' will not improve, and its final course, with ever more frequent military strikes and smothering of political freedom, will dangerously diminish the chances for world-wide endeavors. There never has been a time when it was as imperative that human rights be established in every country as the only means for initiating a general action to save man in the face of a world catrastrophe. In these endeavors the solidarity which has developed among workers groups and humanistic societies is all important. Scientists, also, can play a role since they are familiar with all the aspects and consequences of the contemporary dynamic.

Science and art do not develop only a single idea, a single action, a single goal. In scientific investigation and artistic productions the most diverse possibilities are tried. History, tradition, work, and circumstances have brought it about that the inhabitants of different parts of the world live, desire, and think in different ways. However, the human imagination is sufficiently adaptable and creative that men from under any part of the heavens can understand and feel with one another. This possibility is the source of humanism. The universality of human investigation and creativity goes hand in hand with many-fold realizations. In this sense, contemporary humanism is pluralistic. This however does not mean that every practice is good. An absolute pluralism is just as senseless as a monolithic or completely uniform way of thinking. Both the one and the other destroy thought and creative action itself. Our tolerance would be self-contradictory and untenable if we tolerated violence or lies. But on the other hand, our opposition to evil cannot go so far that it increases hatred and fear.

In some places repressive systems have left such deep emotions of revenge that they can easily explode into destructive eruptions. It is the obligation of every culture to meet the longings of the enslaved, the dispossessed, and the alienated by increasing human understanding, freedom and social rights. Contemporary humanism will not claim that the radical re-creation of the world comes the first of all; it will rather be more concerned with removing the sources of evil (armaments, poverty, frustration, violence) and with creating the conditions for a freer, richer, and fuller life. Today at the climax of history's contradictions, humanism has become the determining appeal of our destiny. While humanists at first were no more than a small European community of Latinists who drew their inspiration and unity from distant and long-ago cultures, the spread of enlightenment and science swelled their ranks throughout the world and brought them closer to everyday life. Not long ago, questions of freedom and the meaning of science were the concern of just a few individuals and universities; but since the end of the Second World War, when the vision of total destruction first appeared the further applications of science, the rules of its pursuit, the overall development of this world became a *to be or not to be* question for every man. Thus humanism itself overflows the narrow corridors of *academe* and begins to run in the veins of contemporary democracy.

University of Zagreb

NOTES

[1] 'Genealogy of Science and Theory of Knowledge', in R. E. Butts and J. Hintikka, eds., *Historical and Philosophical Dimensions of Logic, Methodology and Philosophy of Science*: Part 4 of the Proceedings of the Fifth International Congress of Logic, Methodology and Philosophy of Science, London, Ontario, Canada, 1975, D. Reidel, Dordrecht 1977, pp. 173–183.

[2] According to the *received view*, theories are to be construed as axiomatic calculi in which theoretical terms are given a partial observational interpretation by means of correspondence rules, cf. Frederick Suppe, 'What's Wrong with the Received View of the Structure of Scientific Theories?', *Philosophy of Science* **39** (1972), 1–19. There are two nonlogical vocabularies, first, V_o consisting of directly observable attributes and entities and, second, V_{th} of not directly observable terms. The theory consists of theoretical laws formulated in a language whose all nonlogical terms belong to V_{th}. Correspondence rules contain terms from V_o and V_{th} and are intended to embody various experimental procedures for applying the laws of theory to directly observable phonomena. My concluding remark: the entire scheme breaks down on the impossibility to make a sharp and absolute distinction between observational and theoretical language.

[3] K. Gödel has been led by the positivistic interpretation of empirical sciences to a platonic position: "But, despite their remoteness from sense experience, we do have something like a perception of the objects of set theory, as is seen from the fact that the axioms force themselves upon us as being true. I do not see any reason why we should have less confidence in this kind of perception, i.e., in mathematical intuition, that in sense perception, which induces us to build up physical theories and to expect that future sense perception will agree with them and, moreover, to believe that a question not decidable now has meaning and may be decided in future. ... It by no means follows that ... because [the data of mathematical intuition] cannot be associated with actions of certain things upon our sense organs, [they] are something purely subjective, as Kant asserted. Rather they, too, may represent an aspect of objective reality, but as opposed to the sensations, their presence in us may be due to another kind of relationship between ourselves and reality". K. Gödel, 'What is Cantor's Continuum problem?', *American Mathematical Monthly* **54** (1947) 515–525.

[4] B. W. Bridgman, *The Nature of Physical Theory*, Princeton 1936, rehearses the position

of Plato and Descartes in a more complicated (and confused) form: "Explanation consists merely in analyzing our complicated systems into simpler systems in such a way that we recognize in the complicated system the interplay of elements already so familiar to us that we accept them as not needing explanation". And he argues further: "Since Quantum Theory does not show how the physical systems result from familiar modes of action between familiar constituents, the theory explains nothing". This rejection of Quantum Theory goes beyond Einstein's criticism and has been wide-spread among American philosophers, probably due to the desire to save traditional ontology.

Compare different positions:

Giambattista Vico's motto: *certum quod factum*, an agent can understand fully only what he has himself made.

Kierkegaard: "Concepts, like individuals, have their histories, and are just as incapable of withstanding the ravages of time".

Frege: "If everything were in continual flux, and nothing maintained itself fixed for all time, there would be no longer any possibility of getting to know about the world, and everything would be plunged into confusion .... What is known as the history of concepts is really a history either of our knowledge of concepts or of the meaning of the word'. *The Foundations of Arithmetic*, Oxford, 1950.

Frege: "Often it is only after immense intellectual effort, which may continue over centuries, that humanity at last succeeds in achieving knowledge of a concept in its pure form, by stripping off the irrelevant accretions which veil it from the eye of the mind".

Hegel: "Nothing else will come out from the history but what was already here". In a different context Kant says similarly: "New physics discovered what reason put in nature". According to Hegel, man must, in order to become god, destroy nature and world, and in the end spirit stands alone. "We are in solitude with ourselves". (*Phenomenology of Spirit*). In the plan of the history, absolute idea, through ramification become full aware of all its potentialities.

⁵ Following Boltzmann, V. I. Lenin would write: "Our impressions and concepts are a picture of these things (outside us). Verification of these pictures, distinguishing right ones from wrong ones, is performed by practice". Similarly also Ludwig Wittgenstein took this *picture-theory* from Boltzmann; and so neopositivism and dialectical materialism have the same common root in spite of the exaggerated hostility started by V. I. Lenin's *Materialism and Empiriocriticism*.

⁶ *Consciousness of Philosophical Study of Minds and Machines*, New York, Random House, 1969.

⁷ *Le hasard et la nécessité*. Paris, 1970.

⁸ The conflict among communists concerning the mirror-theory and the autonomy of science and art broke out in Yugoslavia particularly in years 1939 and 1940, leaving more than hundred articles and books about this subject. A reminiscence of this conflict between so-called revisionists and orthodox followers of dialectical materialism (and Stalin's line) has been shaped in my novel published in *Forum*, Yugoslav Academy of Sciences and Arts, 3, 4–5 and 6, 1976, certainly not complete. This struggle around socialist realism and dialectical materialism has been connected with the basic political question in 1938: whether communists have to go together with democratic opposition. The 'hardliners' opposed any alliance with democratic parties while we pre-war 'revisionists' were in favor for such a popular front, like Eurocommunism today.

PATRICK SUPPES

PROBABILISTIC EMPIRICISM
AND RATIONALITY

INTRODUCTION

I shall not define with any precision what I mean by probabilistic empiricism, although it should be clear from the surface meaning that I do want to use probability concepts to deal with metaphysical, epistemological, and ethical matters. I intend deliberately to replace the concept of logical empiricism by that of probabilistic empiricism, and I shall argue that it is probabilistic concepts rather than logical concepts that provide a rich enough framework to justify both our ordinary ways of thinking about the world and our scientific methods of investigation.

In Section I, I examine the impact of probabilistic considerations in contemporary natural science. In Section II, I consider the relation of probability to rational decision making. In Section III, the final section, I sketch some of the problems facing a Bayesian or probabilistic theory of rationality.

I. PROBABILITY IN NATURAL SCIENCE

Probabilistic considerations are a working and natural part of ordinary contemporary science. One of the first uses of probability was in the recognition in the 18th century that an explicit probabilistic theory of errors of measurement was required. When the underlying probability distributions of errors are integrated into the deterministic differential equations of mechanics, for example, the result is a theory of phenomena with randomness. A second source of probability or randomness in actual scientific practice is the random intrusion of effects from outside the system being analyzed. A paradigm case is the intrusion of small amounts of matter and energy from outside the solar system. A third and more fundamental kind of randomness arises when the equations of motion of a phenomenon are themselves stochastic equations. Good physical examples are provided by radioactive decay and, more generally, in the thoroughly probabilistic theory of modern quantum mechanics.

I now go into these ideas in more detail.

171

R. Hilpinen (Ed.), Rationality in Science. 171–190.
Copyright © 1980 *by D. Reidel Publishing Company.*

Measurement Errors in Classical Physics

Already in classical Laplacean astronomy it was recognized that the measurements of initial conditions and boundary conditions are subject to error. When the underlying theoretical probability distributions of these errors are integrated into the deterministic differential equations of mechanics, the result is a theory of phenomena with randomness. There is an inevitable dispersion of accuracy over time, so that intuitively the phenomena in question are less and less *determined* as time progresses. Analyses of a variety of problems in which initial conditions are assumed subject to random variation are important not only in astronomy, but also in engineering mechanics, in chemistry, and in other areas of applied science. Traditionally, the true-blue determinist, however, is not upset by admission that uneliminable errors of measurement produce randomness of initial conditions and, therefore, a nondeterministic analysis of phenomena. The classical response is that the errors of measurement can in principle be eliminated. One of the missions of classical physics was to perfect methods of measurement with the ideal of ultimately eliminating all such errors.

Random External Effects

A second course of randomness in actual scientific practice that does not disturb the true-blue determinist is the random intrusion of effects from outside the system being analyzed. A paradigm case would be the intrusion of small amounts of matter and energy from outside the solar system, or the variations in density of matter in space through which the solar system is moving. The effects of these external intrusions are extraordinarily slight in terms of the motion of the main objects in the solar system, and consequently, even though it is recognized that they exist and that detailed calculation of their behavior is almost certainly out of the question, the attitude is one of serenity because of their relative insignificance.

   An example of a different sort is the following. From the standpoint of the fundamental dynamics of the earth's atmosphere, the science of meteorology treats the earth as a closed system and uses an elaborate set of differential equations to predict the entire motion of the atmosphere. Although it is not really feasible to look at the whole atmosphere this way, it is sometimes considered an approachable ideal. What is not taken into account in the fundamental dynamical equations of meteorology are random disturbances caused by unpredictable events on the surface of the sun. So far as I know, no fundamental analysis of the dynamical equations of the earth's atmosphere attempts to include terms that would account for the appearance of sun spots and other disturbances on the surface of the sun that affect in clearly measurable ways the earth's atmosphere. Even a phenomenon of this sort is not too disturbing to the hard-line determinist, however, and he would be content to remain as firm as Kant in his conviction of the fundamental correctness of Newtonian physics – the fundamental dynamical equations of meteorology are, of course, squarely within the tradition of classical physics.

Stochastic Equations of Motion

A third and more fundamental kind of indeterminism arises when the equations of motion of a phenomenon are themselves stochastic equations. In phenomena subject only to stochastic laws, there is no hope (within the given framework) of predicting or of being able to determine exactly the future or past course of events. Already in this third case, however, there is an important division to be made. Stochastic differential equations can be used as the fundamental dynamical equations of a phenomenon not because of any firm conviction that the phenomena in question are indeterministic in character, but, rather, because of their complexity. Typical instances are these: wave-propagations in inhomogeneous media, chemical systems in which there exist small percentages of impurities, and economic or biological systems whose underlying microanalysis is either unknown or involves too much data collection to make analysis feasible. In the case of biology we can believe that the fundamental physics and chemistry of cells can be understood at the level of classical physics and yet we can apply to the study of the motion of fluids and cells through porous media, etc., stochastic rather than deterministic equations because of the enormous complexity of the situation. The philosopher sitting in detachment away from the agony of the details can still with serenity maintain that the fundamental laws of the universe are deterministic and that all is well with the broad Laplacean or Kantian thesis.

But this is not the end of the story at the level of fundamental theory. I now consider, in sequence, randomness in radioactivity and the more general theory of modern quantum mechanics.

Randomness in Radioactivity

Let us look quickly at some of the historical background to radioactivity. In the autumn of 1895, Roentgen, professor of physics at Wurzburg, accidentally discovered X-rays and, through a beautiful series of experiments, demonstrated their fundamental properties.

The discovery of radioactivity itself followed automatically from experiments connected with phenomena similar to those observed in X-rays. Several investigators found that fluorescent bodies exposed to sunlight gave out a type of radiation similar to that of X-rays. Early in 1896, in trying an experiment of this kind, Henri Becquerel discovered that the specimen of uranium and potassium he was using emitted radiation even in the dark. Shortly thereafter, he found that the emission of radiation by uranium was more or less independent of its state of chemical combination with other substances and that there was no connection between this phenomenon of radiation and that of phosphorescence, the initial subject of his investigations. Also he found that the radiation was more or less independent of the temperature of the uranium compound. Not too much later, radium was discovered by the Curies.

By 1905 at least, physicists were already questioning whether the emission of particles from radioactive substances was deterministic or probabilistic in character. In 1910, Rutherford and Geiger published a paper on this topic that contained an appendix by Bateman on the mathematical properties of Poisson processes, and over the next decade or so, a number of additional studies followed. An excellent review of the literature, including a detailed critique of the various studies, was given by Kohlrausch in 1926. Although methodological criticisms can be made of most of the studies from the standpoint of modern statistics, the weight of the evidence is certainly in favor of the statistical character of radioactive decay. There is no body of systematic evidence that a deterministic law holds, and it seems appropriate to interpret the large number of studies of these matters as directly supporting the thesis that randomness is in nature, and not simply in our ignorance of true causes as Laplace would have wanted it.

Because the case of radioactive decay is an elementary one, I would like to examine in somewhat greater detail its implications for the view that the universe is essentially probabilistic in character or, to put it in more colloquial language, that the world is full of random happenings.

The most important point is to dispel the illusion that because random happenings may be found everywhere, the analysis of phenomena somehow becomes too complex, too disorderly, and consequently too difficult to leave any hope for the development of systematic theory. It is part of my theme that it is a return to realism – realism in the sense of realism of belief and not in the sense of ontological realism – to recognize how schematic any of our knowledge of the universe must be and that it is the character of commonsense knowledge itself to be schematic and probabilistic. What is mistaken and hopeless is to think we can pursue to the bitter end a sequence of determinant events with the whole universe interlocked in one vast deterministic moving forward. Associated with randomness is also independence, and the basic physical assumption of radioactive decay can be stated in a completely elementary and qualitative way in terms of these two concepts (Suppes, 1973), whereas in contrast, if a deterministic theory of the orbits of electrons around nuclei were required, as was once thought in classical physics, the details would be extraordinarily complicated and difficult.

We should note also how natural the tendency is to want to infer complicated effects from the history of the atom and to be psychologically unsatisfied with such a simple assumption. "Surely," we are inclined to say, "something is going on in the atom during all this period, and the effect of these somethings should be cumulative and should distort the probability of decay."

Randomness in Quantum Mechanics

It is undoubtedly true that the instance of radioactive decay can be treated as lacking in theoretical import, because the phenomenon in question and the probabilistic laws of decay that adequately describe the phenomenon do not have extensive theoretical reach; consequently, the possibility very much remains open of deepening theory to account in a deterministic way for the apparently random phenomenon of decay. It is well known that it was Einstein's view that a deeper deterministic theory would be found to subsume quantum mechanics. The search for such theories has come to be known as the search for hidden-variable theories. The term *hidden variables* is picturesquely descriptive of what is desired. The hope is that back of the probabilistic variables observed in quantum mechanics will be found deterministic variables that will account for the observed probabilistic phenomena. A simple example of a classical hidden-variable theory is statistical mechanics. It is part of classical statistical mechanics to postulate a determinant state for individual particles and to introduce randomness in proper Laplacean terms as an expression of our ignorance of the true state.

The analysis of hidden-variable theories has a complicated history in modern physics, beginning, if not earlier, with the celebrated proof of von Neumann that dispersion-free states and, consequently, hidden variables are impossible in quantum mechanics. It is important to realize what the connection between dispersion-free states and hidden variables is. Dispersion-free states correspond intuitively to classical states in which position and momentum, for example, are definitely determined. The central idea is that hidden variables lead to the specification of such dispersion-free states.

The essential assumption of von Neumann's proof is that any linear combination of two quantum mechanical (i.e., Hermitian) operators represents an observable, and the linear combination of expectation values of the operators is the expectation value of the combination. From a conceptual standpoint, this assumption can be criticized as smuggling in unwarranted and rich assumptions about the results of quite different experiments. The point is that three different experiments would ordinarily be required – one for the first observable, a second for the second noncommuting variable, and a third for the linear combination of the two. Given the different experimental configurations required in the three cases, there is no reason to assume that the expectation values will hold for pure states as opposed to quantum mechanical averages. The significance of this argument, stated most clearly, for example, in Bell (1966), is that the results for the linear combination of observables cannot be arrived at simply by computation. If this were the case, matters would be simple. What is important is the requirement that there be a third and distinct experimental arrangement for the measurement of the linear combination.

Various improvements that weaken the assumptions of von Neumann have been made subsequently in the literature.

In a beautiful series of papers beginning with Bell (1964, 1966), a much more reasonable and intuitive treatment of hidden-variable theories has been given, and their impossibility has been demonstrated experimentally at a rather satisfactory level. Without entering into the details, essentially what Bell has been able to show is that, first, if we start with the paradox of Einstein, Podolsky, and Rosen (1935), which argues for the incompleteness of quantum mechanics, and, second, we insist that a hidden-variable theory that removes the incompleteness must satisfy natural conditions of causality and locality, then by considering a simple system with two particles of spin one-half, these conditions cannot be satisfied and there can be no hidden-variable theory for such two-particle systems. Within the context of this analysis, Bell was able to derive an inequality that has come to be known as "Bell's inequality," and this has been used by Clauser, Horne, Shimony, and Holt (1969) and Freedman and Clauser (1972) to show that the existence of local hidden variables imposes restrictions that are, as Bell originally showed, in conflict with quantum mechanics, and that, further, new experimental data are in agreement with quantum mechanics. Moreover, well within the accuracy of experimental error, the data violate the restrictions required by local hidden-variable theories. Recently, Zanotti and I have found a general argument against hidden-variable theories using only negative correlations and the probabilistic concept of exchangeability, introduced many years ago by de Finetti.

I think it is fair to say that the outcome of these various papers, both theoretical and experimental, is to provide perhaps the conceptually most satisfying confirmation of the ultimately statistical character of quantum mechanics that we yet have. Furthermore, from a philosophical standpoint it provides one more blow against Laplacean or Kantian ideas of determinism as regulatory for the behavior of the universe.

II. Probability and Rational Decision Making

For further discussion, I sketch the main ingredients of the Bayesian theory of rationality that has a long tradition, reaching back to Daniel Bernoulli. Some essential features of the theory in standard form are these.

First, the theory is meant to be a theory of rational action and not of rational talk. In ordinary conversation it is certainly true that we attribute rationality both to talk and to action. We are quite prepared to say, "That sounds very rational," in commenting perhaps on a speech by a politician of our own persuasion; at the same time we are also prepared to say of someone that, although he talks very rationally, his actions are most peculiar and at times irrational. Both our commonsense psychology of individuals and the general philosophical view that a man's moral principles are best exhibited by his actions rather than by his words support the notion that the basic thrust of rationality is to be assessed in terms of actions rather than in terms of talk. This thrust is central to the modern Bayesian position.

Second, the theory is designed above all to deal with situations of uncertainty. Both classical moral theory and classical economic theory ignore almost entirely the difficult and subtle problems of making rational decisions in the face of uncertainty. The classical theory of demand in economics, for example, can rely on a purely ordinal theory of value so long as no questions of uncertainty enter. On their own side, traditional moral theorists have simply not concerned themselves with the problem of rational or moral action in the face of uncertainty.

Third, the Bayesian theory requires that it be possible to assign numerical probabilities to beliefs about possible states of affairs, although it is quite consistent with the theory that only qualitative postulates need be accepted. These qualitative postulates should be strong enough to lead to a unique probability distribution on possible states of affairs. The basic qualitative relation is that of one event being judged at least as probable as another.

Fourth, the theory requires that a numerical utility or valuation function be defined for possible consequences of actions taken under various possible states of affairs. The valuation function should be unique up to a linear transformation, that is, unique up to the fixing of an arbitrary unit and an arbitrary zero point. And again, the theory of utility can itself arise from qualitative postulates and need not be directly numerical in character. It is important to emphasize that there is no necessary adoption of hedonism or of utilitarianism based on principles of maximizing pleasure. The evaluation principles can be based on the theory of obligation, and such principles of obligation can enter into the qualitative principles of the theory.

Fifth, the central principle of the theory is to combine the ingredients just stated to lead to the overall single principle of rationality, the principle of maximizing expected utility: One action or decision should be chosen over another if and only if the expected utility of the first is at least as great as that of the second, where expected utility is defined as it has been since the beginning of the 18th century.

The point of the theory as enunciated by Savage (1954) and others is to place axioms on preference among acts in such a way that anyone who satisfies the axioms will automatically satisfy the central principle of maximizing expected utility. This means, among other things, that the way in which the axioms are satisfied will generate a subjective probability distribution on beliefs concerning the true state of nature and a utility function on the set of consequences arising from any possible action and any possible state of nature. The expectation of an act is then taken with respect to the subjective probability distribution of beliefs about the states of nature and the utility function on the set of consequences.

Savage's theory requires only seven axioms, but their formulation is rather awkward and formally somewhat complicated. Even though I have some preference for a later formulation of my own (Suppes, 1956), because of their prominent place in the literature let me briefly summarize Savage's axioms as a point of reference for later discussion. Axiom 1 requires that the preference among acts be transitive and, given any two acts, that one is at least weakly preferred to the other. Axiom 2 extends this ordering assumption to having the same property hold when the domain of the definition of acts is restricted to some given set of states of nature. Axiom 3 requires that knowledge of an event cannot change preferences among consequences. Axiom 4 requires that given any two sets of states of nature, that is, any two events, one is at least as probable as the other; in other words, qualitative probability among events is strongly connected. Axiom 5 excludes the trivial case in which all consequences are equivalent in utility. Axiom 6 is a strong partitioning axiom closely related to an earlier axiom of de Finetti and Koopman: If event A is less probable than event B, then there is a partition of the states of nature such that the union of each element of the partition with A is less probable than B. Finally, Axiom 7 is the formulation of the sure-thing principle, which says that if one act always has outcomes at least as good as the second has, and on some occasions better outcomes, then it should be always chosen over the second.

To return to the earlier contrast between rationality of talk and rationality of action, we must give further consideration to one point about the Bayesian approach to rationality. On the one hand, there is a natural tendency to think of Bayesian methods as an extension of those of classical logic. On the other hand, there is a departure from classical logic in the consideration of actions and events rather than sentences or propositions. This sharp dichotomy should be greeted with suspicion, because of the lack of a sharp distinction between deductive and inductive reasoning in ordinary talk. In ordinary argumentation and in the use of words that state evidential grounds, we never make a clear and definite distinction between a deductive and an inductive consideration.

It seems peculiar, therefore, to have the theory of rationality somehow draw a sharp distinction between actions on the one hand and sentences on the other, assigning the one to the theory of rationality as such and the other to logic.

I think this sharp dichotomy is more a matter of appearances than of reality. First, there is a classical tradition of ambiguity in probability theory of whether to speak of the probability of events or of propositions, and I do not think it a matter of great consequence which one we choose. Second, the assignment of values to consequences – consequences in the sense of the consequences of action, of course, and not logical consequences – is simply the addition of a further feature not considered in classical logical terms. If we restrict Bayesian theory or the general theory of rationality in situations of uncertainty to the theory of partial belief, I think we can claim that the theory of partial belief or the theory of subjective probability – call it what we will – is a natural and continuous extension of classical deductive canons of inference.

Notice that when the true state of affairs is known, then as already remarked in the classical economic theory of demand, instead of a theory of partial belief we need only a method for knowing ordinal preferences and for selecting the act that leads to the ordinally best overall consequence. Logic could of course enter into evaluating consequences, but the important point is that a direct role for the logic of belief is not required. In one straightforward and correct sense, the logic of belief is trivial when the true state of nature is known. In that sense, the inductive logic of subjective probability is a clear extension of classical deductive logic, and the character of this extension in general terms is clearly recognized by everyone.

III. PROBLEMS FOR BAYESIANS

It might be thought that the kind of Bayesian theory I have sketched leads to a happy paradise for probabilistic empiricists of my ilk and one that I would be happy to reside in without quarrel, doubt, or aspiration for change. But it is characteristic of the kind of philosophy I believe in not to be content with a theory so simple and apparently so complete as the Bayesian one.

Thus I turn to three kinds of problems that Bayesian theories of rationality face and indicate in part my own approach to meeting these problems. The first concerns the use of structural axioms that are not axioms of pure rationality; the second, problems of inexact measurement; and the third, problems of extension of the framework of concepts as new and surprising situations are encountered.

The Problem of Structural Axioms

At the time of the Third Berkeley Symposium on Probability and Statistics in 1955, I introduced the distinction between structure axioms and rationality axioms in the theory of rational decision making (Suppes, 1956). Intuitively, a structure axiom, as opposed to an axiom of pure rationality, requires that some special features of the environment be present and that they not impose a constraint on the rationality of the decision maker that must be satisfied always and everywhere. In most cases, structure axioms are existential in character, but if defined notions are introduced in the formulation of axioms, then it is possible for the axioms to appear universal in character, but still to express structural conditions. Many examples of this kind may be found in my article on finite equal-interval measurement structures (Suppes, 1972). A structure axiom is usually transparently so when the axioms are expressed in terms of the primitive concepts of the theory, which merely provides reinforcement for Tarski's admonition that we only understand the complexity of a given set of axioms when they are formulated in terms of the primitive concepts. In the case of Savage's seven postulates intuitively described already, two – the one concerning the existence of consequences that are not equivalent and the powerful partitioning axiom – are structure axioms, and they are obviously existential in character.

Savage defended his strong partitioning axiom (Axiom 6) by holding it applicable to a situation in which there is a coin that the decision maker believes is fair for any finite sequence of flips. Of the several objections to this argument, the principal one, it seems to me, is the strong finitistic one that we can scarcely wait through an indefinitely large number of flips in order to determine what decision to make. The exactness of Savage's results depends upon unbounded finite sequences of flips, and no effort is made in his theoretical analysis to deal with the case of a strictly bounded sequence as we shall in the sequel here.

Because the introduction of something corresponding to a fixed finite sequence of flips is central to the axioms I shall consider later, it is important to be clear on the point at issue. I think a challenge can be put to the introduction of any sort of device like that of a fair coin, but certainly the challenge must be more restricted in nature if the sequence of flips is strictly bounded, for example, if it is understood that we need to flip a coin only seven or eight times to get the appropriate fine structure into the decision situation. In almost all environments in which we are forced to choose an act, the set of acts or decisions open to us is small, and the events that we consider relevant are small in number.

The same air of unrealism prevalent in some of the traditional ideal-izations in ethics results from insisting on an enlarged decision frame-work of the kind that is the basis of Savage's theory. In fact, the grandiose way of formulating the Savage-type theory, in which we are to think of consequences as future histories of the universe, seems to me wholly unsatisfactory. There is a tendency to formulate the theory on a grand scale and to invest it with a kind of aura of completeness that it actually does not have. On the other hand, when the theory is restricted to small worlds, to use Savage's elegant phrase, there is some practical hope of applying the theory and using it as a guide for action.

In a preliminary fashion I have formulated the description of struc-ture axioms in terms of existential requirements, that is, existential re-quirements that must be satisfied by the environment and that we ordinarily do not think of as part of the mental machinery of the ra-tional decision maker. A more general way of characterizing axioms of pure rationality is to require that they be closed under submodels on the reasonable assumption that this restriction to submodels is one that may occur in practice. Closure under submodels is a restriction on the environment, not on the mental apparatus of the decision maker.

To keep the technical apparatus simple, let us restrict ourselves to a basic set S of states of nature and a binary ordering relation $\geq$ of qualitative probability on subsets of S with the usual Boolean opera-tions of union, intersection, and complementation having their intuitive meaning in terms of events. This restriction would seem to reduce wholesale the theory of rational action to the theory of rational belief, but the restriction serves only technical purposes. In Savage's proof and in others like it, the first thing to establish is that the beliefs about the state of nature lead to construction of a probability distribution express-ing these beliefs, and we may assume that the restriction now being contemplated deals with the first part of the subject. There is also an independent interest in the theory of rational belief and, if one wishes, the axioms can be so regarded as providing a self-contained treatment, although personally I prefer to combine it into one whole considered as the theory of rational action.

Given the structures just described, I then say that an axiom about such structures is an axiom of pure rationality only if it is closed under submodels. More explicitly, closure under submodels means that if the axiom is satisfied for a pair $\langle S, \geq \rangle$, then it is satisfied for any nonempty subset of S with the binary relation $\geq$ restricted to the power set of the subset, that is, restricted to the set of all subsets of the given subset. Clearly, this condition of closure under submodels is close to the axiom on the independence of irrelevant alternatives in the theory of choice. I remind you that this latter axiom says that if we express a preference among candidates for office, and if one candidate is removed from the list by death or other reasons, then our preference ordering of the remaining candidates should be unchanged. This axiom satisfies closure under submodels and is the paradigm of what is intended by the condition.

A different, but related, way of defining axioms of pure rationality is that such an axiom must be a necessary consequence of the existence of the intended numerical representation, in the case under discussion, the numerical representation in terms of subjective probability and of numerical utility. Thus, for example, it is easy to show that Savage's sixth axiom, which is not closed under submodels, is also not a consequence of the intended numerical representation leading to the principle of maximizing expected utility.

Note that the criterion of necessary consequence of the intended numerical representation is a necessary and sufficient condition, but an extrinsic one. The criterion of closure under submodels is only necessary but, on the other hand, it is intrinsic in character. So far as I can see, we have at the present time no good method of attack on the problem of finding an intrinsic necessary and sufficient characterization of axioms of pure rationality.

I linger upon this point, because it is one of the few areas in which we can actually have a conceptually structured discussion of what one might mean by rationality and of what general criteria axioms of rationality should satisfy.

Inexact Measurement

Almost everyone who reflects on the problems of measuring beliefs in the tradition of subjective probability feels some uneasiness about asking for arbitrary refinement of the measurement. There is something in fact paradoxical about the thrust of Savage's theory and others for an exact probability distribution on the states of nature. If we compare this situation with the situation that exists in the case of physical measurement, we can see how absurd the striving for exact results is. We would consider it unthinkable to require of a physical theory that it be based on or that it obtain exact measurements of mass or position. In all cases, we expect a statement about the nature of errors of measurement, and in standard practice, we expect some estimate of the error of the measurements. It is fair to say that the explicit realization that errors of observation arise and yet that an explicit theory of these errors can be given is one of the main contributions of the theory of probability to general scientific methodology. Without such an explicit theory of error, much of the more exact and sophisticated science of the latter part of the 19th century and of the 20th century would have been conceptually difficult, if not impossible, in terms of serious confrontation between theory and data.

On the other hand, if we are asked for our beliefs about the probability of rain tomorrow or the rate of inflation or the growth of the gross national product, we find it hard to think through how to assign definite errors of measurement to the process of determining these beliefs. In fact, so far as I know, there is no serious literature on the subject of developing anything like a direct analogy with the theory of errors of observation in measurement in the experimental sciences. However, another approach that has been discussed much less than one might think seems promising. This is the approach that assigns not exact probabilities, but upper and lower probabilities, to the occurrence of an event. The general idea of such an approach has already been stated in Bayes' memoir (1763). Although Bayes put the matter in somewhat different language, his meaning is clear:

From the preceding proposition it is plain, that in the case of such an event as I there call M, from the number of times it happens and fails in a certain number of trials, without knowing anything more concerning it, one may give a guess whereabouts its probability is, and, by the usual methods computing the magnitudes of the areas there mentioned, see the chance that the guess is right. (p. 392).

The explicit theory, however, of upper and lower probabilities as upper and lower estimates of the probability of the occurrence of events is quite recent. The main prior publications are Smith (1961), Good (1962), and Dempster (1967). The simple axiomatic treatment that I give has not previously been considered in the literature, except in my own prior work (Suppes, 1974). My axioms for such inexact measurement of belief are not the most general axioms possible, but they are, I think, simple and rather elegant in character.

From a formal standpoint, the basic structures to which the axioms apply are quadruples $\langle X, \mathscr{F}, \mathscr{S}, \geq \rangle$, where X is a nonempty set, $\mathscr{F}$ is an algebra of subsets of X, that is, $\mathscr{F}$ is a nonempty family of subsets of X and is closed under union and complementation, $\mathscr{S}$ is a similar algebra of sets, intuitively the events that are used for standard measurements, and I shall refer to the events in $\mathscr{S}$ as *standard* events S, T, etc. The relation $\geq$ is the familiar ordering relation on $\mathscr{F}$. I use standard abbreviations for equivalence and strict ordering in terms of the weak ordering relation. (A weak ordering is transitive and strongly connected; that is, for any events A and B, either $A \geq B$ or $B \geq A$.)

DEFINITION. *A structure* $\mathscr{X} = \langle X, \mathscr{F}, \mathscr{S}, \geq \rangle$ *is a* finite approximate measurement structure for beliefs *if and only if X is a nonempty set, $\mathscr{F}$ and $\mathscr{S}$ are algebras of sets of X, and the following axioms are satisfied for every A, B, and C in $\mathscr{F}$ and every S and T in $\mathscr{S}$:*

AXIOM 1. *The relation $\geq$ is a weak ordering of $\mathscr{F}$;*

AXIOM 2. *If $A \cap C = \varnothing$ and $B \cap C = \varnothing$, then $A \geq B$ if and only if $A \cup C \geq B \cup C$;*

AXIOM 3. *$A \geq \varnothing$;*

AXIOM 4. *$X > \varnothing$;*

AXIOM 5. *$\mathscr{S}$ is a finite subset of $\mathscr{F}$;*

AXIOM 6. *If $S \neq \varnothing$, then $S > \varnothing$;*

AXIOM 7. *If $S \geq T$, then there is a V in $\mathscr{S}$ such that $S \approx T \cup V$.*

In comparing Axioms 3 and 6, note that A is an arbitrary element of the general algebra $\mathscr{F}$, but event S (referred to in Axiom 6) is an arbitrary element of the subalgebra $\mathscr{S}$. Also in Axiom 7, S and T are standard events in the subalgebra $\mathscr{S}$, not arbitrary events in the general algebra. Axioms 1 to 4 are just the familiar de Finetti axioms without any change. Because all the standard events (finite in number) are also events (Axiom 5), Axioms 1 to 4 hold for standard events as well as for arbitrary events. Axiom 6 guarantees that every minimal element of the subalgebra $\mathcal{S}$ has positive qualitative probability. Technically a minimal element of $\mathcal{S}$ is any event A in $\mathcal{S}$ such that $A \neq \varnothing$, and it is not the case that there is a nonempty B in $\mathcal{S}$ such that B is a proper subset of A. A *minimal open interval* (S, S') of $\mathcal{S}$ is such that $S < S'$ and $S' - S$ is equivalent to a minimal element of $\mathcal{S}$. Axiom 7 is the main structural axiom, which holds only for the subalgebra and not for the general algebra; it formulates an extremely simple solvability condition for standard events. It was stated in this form in Suppes (1969, p. 6) but in this earlier case for the general algebra $\mathcal{F}$.

Because upper and lower probabilities are not widely used, it will be helpful to state the properties we expect them to satisfy. The important three are the following, of which the third expresses the analog of the additivity of ordinary probability:

I. $P_*(A) \geqslant 0$.
II. $P_*(X) = P^*(X) = 1$.
III.   If $A \cap B = \varnothing$,   then
$$P_*(A) + P_*(B) \leqslant P_*(A \cup B) \leqslant P_*(A) + P^*(B)$$
$$\leqslant P^*(A \cup B) \leqslant P^*(A) + P^*(B).$$

For standard events the upper and lower probabilities will be identical and equal to the exact probability, but for an arbitrary event the upper and lower probabilities provide bounds, and I think of the true probability of such an event as lying in the open interval given by the upper and lower probabilities.

The theorem that we can prove about such inexact measurement is the following. The proof is given in Suppes (1974).

THEOREM. *Let* $\mathcal{X} = \langle X, \mathcal{F}, \mathcal{S}, \geqslant \rangle$ *be a finite approximate measurement structure for beliefs. Then*

(i) *there exists a probability measure* P *on* $\mathcal{S}$ *such that for any two standard events* S *and* T

$S \geqslant T$ *if and only if* $P(S) \geqslant P(T)$,

(ii) *the measure* P *is unique and assigns the same positive probability to each minimal event of* $\mathcal{S}$,

(iii) *if we define* P_* *and* P^* *as follows:*
     (a) *for any event* A *in* $\mathcal{F}$ *equivalent to some standard event* S,

$$P_*(A) = P^*(A) = P(S),$$

(b) *for any A in $\mathscr{F}$ not equivalent to some standard event S, but lying in the minimal open interval (S, S') for standard events S and S',*

$$P_*(A) = P(S) \text{ and } P^*(A) = P(S'),$$

then P_* and P^* satisfy conditions I to III for upper and lower probabilities on $\mathscr{F}$, and
(c) *if n is the number of minimal elements in $\mathscr{S}$, then for every A in $\mathscr{F}$*

$$P^*(A) - P_*(A) \leqslant 1/n.$$

A further problem of precision in measurement that still needs to be dealt with is the common response to upper and lower probabilities in the theory of rational belief or behavior, which is, "Well, the problem of exact measurement of probability has now been replaced by the exact measurement of upper and lower probabilities." A clear and definite response can be given to this criticism, and it is the following. If there are n minimal standard events, then the probabilities of the 2^n standard events are known exactly as rational numbers of the form m/n. It is true that these rational numbers are given exactly, but any additional precision takes us outside the axiomatic framework. The upper and lower probabilities are all known in terms of this selected fixed finite set of rational numbers, and, at the very least, we have minimized the claims about precision by restricting ourselves to this fixed set. It should be apparent that there is a close analogy in this situation to the one that exists in the use of an equal-arm balance in the measurement of mass or in the use of a measuring rod in the measurement of distance. For the given scales of measurement, uniqueness is expected for the standard events and for the equal-arm balance the standard weights, but precision beyond the selected fixed finite set of standard weights is not expected and cannot be obtained. In fact, as we move to a finer and better balance, we expect to show that we do not have precision for the fixed finite set of weights used in the grosser and less refined scale. In the same way, we could prove a theorem about refinements of the upper and lower probabilities assigned to events by increasing the number of minimal events, but these technical considerations will not be entered into here.

From the standpoint of the problems of exactness of measurement and the related difficulties I have discussed, I think that the axioms embodied in the characterization of approximate measurement structures for beliefs given above do answer some of the main problems we encounter. They provide in a form that is surprisingly simple a semirealistic theory of how we can measure beliefs approximately, and they impose axioms of approximation on the rational decision maker. But, in my view, they do not provide a complete solution to the problem of measurement.

Certainly one of the most difficult aspects we quickly encounter in actual experimentation or actual attempts to measure the beliefs of individuals is the extent to which they are influenced by context, and no considerations of context have been entered into. It might be thought that the rational man should be exempt from context but, as I have indicated elsewhere, such a view of rationality is to me as unrealistic as the claim that the perfect theory of the motion of bodies should consider motion only in a perfect vacuum and not in a fluid or gaseous environment. It is also absurd, it seems to me, to have a precise theory for such a simple matter as that of motion in a gas or a liquid and not to have a similar need and similar desire for such a theory of context in the case of rational belief or rational action. I am not, on the other hand, prepared to make a serious extension in this direction in the present paper. I simply make the modest claim that the axioms I have introduced do represent a constructive forward step to solving the problem of providing a more realistic theory of rationality, as far as problems of measurement are concerned.

Rational Problem Solving and Conceptual Innovation

A deeper difficulty for Bayesian conceptions of rationality is the almost complete inadequacy of the account given of appropriate approaches to new situations whose features are not strongly structured in terms of past experience. It is important to emphasize that both the tradition of moral philosophy and the more recent tradition of rational decision making are in many respects traditions of conservativism.

This topic also brings us back around to the issue of rational talk. To a large extent, we find our way in a new situation of any complexity by talking about it. It is a feature of good talk that it is rational in examining how we should explore a new situation and solve the problems we encounter there. Although the theory of rational talk, as already indicated, has scarcely been developed, it is clear that in confronting new situations we can all make quite good judgments about the rationality of the talk of someone else who is with us, even if we are rather poor in judging our own talk. It seems to me an important matter for the general theory of rationality to try to clarify the criteria that such talk satisfies.

One final point I want to make in this matter, however, is that rational talk is not the expression of determinate and fixed intentions but itself occurs in an environment of internal flux of phenomena. The detailed grammar, or the detailed semantics of what we say, will not be thought out in a determinate way and will not be known to us in general before we make a given comment, however rational it may be. The flux of our internal mental machinery and the flux of the stimuli outside our minds had best be thought of in random terms, and it is only the global averaging reflected in our talk or our actions that properly should be regarded as rational. Rationality is a concept like temperature: It has no meaning when pushed downward to too small a scale of phenomena.

Stanford University

REFERENCES

Bayes, T. 'An essay toward solving a problem in the doctrine of chance.' *Philosophical Transactions of the Royal Society*, 1763, **58**, 370–418.

Bell, J. S. 'On the Einstein Podolsky Rosen paradox', *Physics*, 1964, 1, 195–200.

Bell, J. S. 'On the problem of hidden variables in quantum mechanics', *Reviews of Modern Physics*, 1966, **38**, 447–452.

Clauser, J. F., Horne, M. A., Shimony, A., & Holt, R. A. 'Proposed experiment to test local hidden-variable theories', *Physical Review Letters*, 1969, **23**, 880–884.

Dempster, A. P. 'Upper and lower probabilities induced by a multivalued mapping', *Annals of Mathematical Statistics*, 1967, **38**, 325–340.

Einstein, A., Podolsky, B., & Rosen, N. 'Can quantum-mechanical description of physical reality be considered complete?', *Physical Review*, 1935, **47**, 777–780.

Freedman, S. J., & Clauser, J. F. 'Experimental test of local hidden-variable theories', *Physical Review Letters*, 1972, **28**, 938–941.

Good, I. J. 'Subjective probability as the measure of a non-measurable set', In E. Nagel, P. Suppes, and A. Tarski (Eds.), *Logic, methodology and philosophy of science*. Stanford, Calif.: Stanford University Press, 1962.

Kohlrausch, K. W. F. 'Der experimentelle Beweis für den statistischen Charakter des radioaktiven Zerfallsgesetzes', *Ergebnisse der Exakten Naturwissenschaften*, 1926, **5**, 192–212.

Rutherford, E., & Geiger, H. 'The probability variations in the distribution of α particles', *London, Edinburgh, and Dublin Philosophical Magazine*, 1910, **20**, 698–707.

Savage, L. J. *The foundations of statistics*. New York: Wiley, 1954.

Smith, C. A. B. 'Consistency in statistical inference and decision', *Journal of the Royal Statistical Society* (Series B), 1961, **23**, 1–25.

Suppes, P. 'The role of subjective probability and utility in decision-making', *Proceedings of the Third Berkeley Symposium on Mathematical Statistics and Probability*, 1956, **5**, 61–73.

Suppes, P. *Studies in the methodology and foundations of science: Selected papers from 1951 to 1969*. Dordrecht: Reidel, 1969.

Suppes, P. 'Finite equal-interval measurement structures', *Theoria*, 1972, **38**, 45–63.

Suppes, P. 'New Foundations of objective probability: Axioms for propensities.' In P. Suppes, L. Henkin, G. C. Moisil, and A. A. Sosa (Eds.), *Proceedings of the Fourth International Congress for Logic, Methodology and Philosophy of Science*, Bucharest, 1971. Amsterdam: North-Holland, 1973.

Suppes, P. 'The measurement of belief', *Journal of the Royal Statistical Society* (Series B), 1974, **36**, 160–175.

KNUT ERIK TRANØY

NORMS OF INQUIRY: RATIONALITY, CONSISTENCY REQUIREMENTS AND NORMATIVE CONFLICT

1.

In two earlier papers,[1] I have presented and discussed the ideology of science, "science" taken in the broad sense of systematic and organized cognitive inquiry. By "the ideology of science", I understand the norms and values presupposed in the conduct of inquiry. This ideology I take to be a normative system: a finite and ordered set of norms and values. Its function is, briefly, to guide (to steer) and to legitimate (to justify) decisions and actions taken in the course of inquiry. I take it, moreover, that such "normative activity" of steering and justifying inquiry cannot be effected without appeal to norms and values.

Methodologies may be considered as sub-sets within the ideology of science. If I may take for granted a distinction between internal and external norms of inquiry, methodologies naturally stand out as embodying above all the internal norms and values of science.

In previous papers I have discussed, *inter alia*, problems concerning the contents and structure as well as the mode of existence of the ideology of science, from a historical and a systematic point of view. In the present paper I shall consider problems connected with the idea of regarding such ideologies, and methodologies in the first place, as normative *systems*.

The property of being systematic may itself be thought of as one of the normative requirements which must be met by any body of knowledge and knowledge claims expressed in language. Systematicity is itself a methodological norm, though one which is by no means clear or well defined outside specific contexts of formal logic or systems theory. Now, we do apply or appeal to ideals which are naturally indicated by words such as, "system" and "systematic" also, and primarily, outside the more specialized uses given to these terms in branches of formal logic. Requirements of systematicity are related to a desire for *order* as opposed to disorder, confusion, and chaos. They are at the heart of ideals of rationality as we like to think such ideals actualized in our search for knowledge as well as in our linguistic expressions of what we find.

191

R. Hilpinen (Ed.), Rationality in Science. 191–202.
Copyright © 1980 by D. Reidel Publishing Company.

The problem dealt with in the present paper is roughly this. There is a connection between justification and rationality in science. Justifiability may be a prerequisite of rationality, and a consistent ideology of science may be a prerequisite of justifiability. Hence the importance of consistency requirements applied to normative systems.

Before I proceed, I should say that I am not a logician; and perhaps one should leave problems of consistency requirements and such like to those properly qualified to deal with them. I also feel, however, that such problems should not be the exclusive concern of special competence and interests. One of the notions behind this paper is the hunch or suspicion that there are certain intriguing problems in the border area between ethics, logic and philosophy of science which are generally overlooked (or so it seems to me), and which can perhaps be brought out by a closer study of certain features of methodologies viewed as normative systems.

2 .

Consistency is a basic methodological norm. It is a minimum requirement of rationality in science and, to some extent, in conduct in general as well. But the notion of consistency is hardly well defined, not even in science, and certainly not for my special concern here: for normative systems. Even where it is well defined, as in branches of logic, it is not a sufficient condition of rationality, and not of systematicity either.

"A logistic system is consistent if there is no theorem whose negation is a theorem."[2] The set of *all* true propositions, however, is presumably consistent in this sense; but would we call it a system? In the case of logistic systems there are a series of other requirements which have to be fulfilled. It is obvious that these requirements are much too strong for systems outside of formal logic. Think, for instance, of the periodic system in chemistry or classificatory systems in botany and zoology.

What we need, then, is something more than consistency and less than formal deducibility and provability. What may then come to mind are other, less rigorous and general methodological ideals or requirements which also have important functions in formal logic. To achieve some measure of *completeness*, we try to include in the system that and only that which "belongs" or is "relevant". The notion of *independence*, rigorously defined for axiom sets, is related to less rigorous notions of redundancy and economy: Ockham's razor, in other words. Already, the link with ideals of *testability* and *controllability* is obvious.

My point in making these remarks is the following: the consistency requirement is usually found side by side with other methodological requirements. Consistency requirements never walk alone. Methodological norms never do. They support each other in our attempts to arrive at systems of knowledge (claims) which are delimited, ordered, and coherent. However vague and hard to define, *coherence* may be a key term for that which is more than mere non-contradiction and less than a deductive system.

Some such intermediate requirement must be applicable also to normative systems. A requirement of non-contradiction alone includes *all* logically consistent norms and values in one normative system, which is not helpful for our purposes. In ethics, as in logic and in empirical science, we are, indeed, familiar with distinct systems which turn out to be incompatible with each other without necessarily contradicting each other (unless we define "incompatible" to entail contradiction).[3] And with distinct systems which may "cooperate" or have interlocking functions without being, in a strict sense, logical consequences of each other.

Indeed "a general theory of scientific rationality" (as stated in the description of the research project), requires that the internal and the external norms and values of the ideology of science – the concerns dictated by considerations of human welfare, and those dictated by the methodological needs of scientific inquiry – can be made to cohere and function together in an orderly and consistent manner, since it is also quite conceivable that they should be incompatible or even inconsistent. To contribute to this may be seen as a joint task for a rational science policy and philosophy of science.

3.

By a norm I mean a proposition or statement which essentially contains a permission, an obligation or a prohibition.

By a value I mean something good or bad. To avoid being detained by the notorious difficulties of an adequate concept of value, I shall here only point to examples of values. Need fulfilment is one type of value, crucial in human welfare. The attainment of desired and chosen ends of action is another type of value. Justice in the basic social structure is a central moral and political value. The experience of something as better than (richer, more joyful, more pleasant than) something else is also valuable. (I have not thereby said that it is good.) Suffering and misery are negative values; and so are injustice, and frustration of needs and the experience of something as worse (poorer, more painful) than something else. The diversity of values is overwhelming, no matter what concept or theory of value we adopt.

A normative system is, in my usage, a finite and ordered set of norms and values.

Relations must obtain between norms and values; and again I must move with caution; the field is unclear and controversial. I shall make one or two assumptions only. One is that values are sometimes basic to norms – in which case norms are derivable from values in senses to be specified later. If, for a given field of activity, we can identify a basic end (goal, aim or purpose), then the attainment of that end is a basic value in the normative system which serves to guide and legitimate activity in that field.

Norms can be "derived" from this value either by means-end relations, or by formal logical relations. Examples will be given below.

Methodology is concerned with systematic cognitive activity. The end and the basic value of methology as a normative system, is, I shall say, to optimize the truth-output,[4] the knowledge and insight resulting from cognitive inquiry. It is tempting to think of this value as the prime and dominant value of methodology. In the words of Rawls: "Being first virtues of human activities truth and justice are uncompromising".[5] This could be taken to mean: if other, subordinate values "clash" with truth, then the others have to yield. Truth always takes precedence; it is ranked as a *higher* value than any other value in the system of methodology. (The glibness of such formulations conceals their lack of precision.) This ranking, plus the fact that there is only one highest value might make it possible to avoid value conflict – values contradicting each other – in the system. If the *norms* of the system are either logically derived from the one basic value or means-end related to it, it might seem reasonable to expect that the norms of the system would also be consistent.

I confess that I have thought in this way, and that I have had to change my mind.

4.

Take now the normative system of morality in general. Two features stand out. It is a many-valued system with no clear hierarchical ordering of its values. Secondly, it offers notorious examples of values which are sometimes and unavoidably incompatible. The demands of justice are not always compatible with the demands of mercy; those of honesty clash with those of kindness. It may, in some cases, turn out to be *impossible* to find a course of action which is both just and merciful, or honest and not unkind. I shall not attempt to decide whether or not this impossibility should be called logical. Primarily it does not seem to be a clash between propositions, statements or sentences. But perhaps between possibilities; there are situations in which mercy and justice are not com-possible. Do we say or should we say that, therefore, the system of morality is inconsistent?

I suggested (end of Section 3) that a normative system which is one-valued and well-ordered might conceivably be without value conflicts of the kind exemplified by mercy and justice. More specifically, it might be thought that general methodology might be conflict free in this sense. But it is not.

Central and highly ranked values in any methodology of inquiry are *creativity and originality*. Another indispensable value is indicated by the terms *intersubjectivity* and *consensus*. But it seems impossible to be original and creative in science without departing from and breaking with an established consensus. This impossibility (necessity) appears to be conceptual or logical, too. As is perhaps also the opposition between mercy and justice.

Conceptual or not, it is plain to see that we cannot treat this kind of conflict as a contradiction. For where we find a contradiction, we re-establish order by one of two devices: we reject one of the two offending parties from the system, or we reject another element in the system which generated the contradiction.

Clearly we cannot save any system of methodology by rejecting from it either intersubjectivity or creativity and it is probably clear that that which generates the need for both of them is the basic value of the system: optimizing the knowledge output of inquiry. Nor do we improve the system of morality by trying to do without either mercy or justice. A moral system which completely lacked either is immoral. In the case of a reprieve we do not reject justice as a value; we simply decide, in the case at hand, to put mercy higher. We do so, one might say, precisely with an awareness of the absolute indispensability for the system of *both* values.

For the soundness of scientific inquiry, both consensus and creativity are essential, although we can foresee conflict between them as unavoidable. The history and sociology of science contain ample evidence for the empirical manifestations of such value conflict: resistance and skepticism on the part of the reigning establishment vis à vis new ideas; the loneliness and obscurity of some of the greatest (Mendel); the bandwagon effect of successful novelties; the security offered by the consensus of "normal science" versus the risky and uncertain promises of creative originality.

However, there is nothing here which invites the conclusion that methodologies with such conflicting values are inconsistent. Consensus and creativity do not always conflict. Progress in science depends precisely on their successful interplay. In the attempt to gain a better understanding of science and its commitments, an awareness of such features is probably desirable. It may help us explain some of the more baffling features of science as a social system, perhaps even to handle them better. There will probably be other such conflicting values. (In the discussion Keith Lehrer mentioned the conflict between maximizing content and staying close to data.)

5.

One of the disturbing features of the approach so far adopted is, I feel, the lack of clarity in one of the basic distinctions: that between norm and value. In principle it seems possible to make the distinction reasonably clear. Norms, I have said, are naturally expressed in the language of deontic logic. Values require terms such as good and bad, better and worse, needs, ends and purposes.

We also know that the logic of norms is more tractable than the logic of values. The logic linking norms and values is more problematic than the "internal" logic of norms (deontic logic). None of this facilitates a clear formulation of questions of consistency and, consequently, of rationality applied to normative systems.

So the program implicit in the distinction between norms and values is difficult to implement; and I have not succeeded in doing so either. I suggested (Section 3) that methodological *norms* can be "derived" from values either by more formal modes of reasoning or by means-end considerations. Thus, if "optimizing the truth-output" (or the growth of knowledge, to use a current phase) is a basic value of inquiry, we seem to be in some sense logically committeed to certain norms concerning the acceptance of true (well confirmed) and the rejection of false (inadequately supported) propositions.[6] At the same time it seems clear that, given such norms and purposes, both consensus and creativity as well as a host of other methodological ideals or requirements can be seen as instrumental, and their instrumentality is precisely one of our reasons for sticking to them. However, in using words such as "methodological ideals or requirements", I (deliberately) avoid calling them either norms or values, or virtues, for that matter. Clearly, both norms and values and virtues can be considered and judged as instruments and very naturally so if we view things in a teleological perspective, which seems both legitimate and necessary in the case of methodologies. A methodology is not an "end in itself".

I cannot offer arguments to show that this lack of clarity in the concepts of norm and value is to be regretted. Perhaps the often very loose language which we use in talking about these matters - referring to methodological ideals or notions by single words such as "consensus", "creativity", "consistency", "completeness", etc. - is not objectionable after all; it may be in order as it is. (The pattern was set - or followed - already in Merton's classical papers[7] on what I call the norms of inquiry. His CUDOS-norms are simply labelled Communism, Universalism, Disinterestedness, and Organized Skepticism.) There may be nothing to be gained by seeking more explicit and definitive formulations.

Yet I think we may again profit from a look at other normative systems. Take, for instance, that of morality in general, or, to choose a more specific example, that of medical ethics. One might say that parts of these systems, mainly some of their *norms*, in a fairly strict sense of this word, are expressed with considerable explicitness and rigor in written and formally adopted *codes* or collections of rules. The positive laws, the codified legal system, of a nation or community, can be seen as such partial and explicit formalization of the general moral system of the community in question. The various national and international biomedical declarations and codes of ethics of the medical profession may be viewed in a similar light.

We can also, I suppose, see good reasons for such formal codification of norms. It is a necessary step when a certain kind of *institutionalization* is desired or required. It may be justifiable precisely in the interests of law and order. This stands to reason: for to the extent that we are able to express ourselves and our commitments in the language of norms - of deontic logic – we can also more successfully secure the attainment of, for instance, such things as consistency requirements, clarity, systematicity. In deontic language, we can more readily decide what the logical consequences are of given norms, and what premisses are necessary to derive or support a desired normative conclusion. Not as perfectly as in logistic systems, but more orderly than in the world and language of values. Codified sets of formal and carefully formulated rules may also serve the purpose of minimizing the undesirable effects of potential value conflicts in many-valued normative systems, and may positively contribute, instead, to the peaceful coexistence of indispensable values which do not always automatically function in harmonious concert with each other.

6.

If a codification of the norms of a normative system can contribute to a reduction of conflicts which the system is otherwise likely to generate or unable to prevent, why should we not also attempt a more formal codification of our rules of methodology?

To put the matter crudely: because it might presuppose both a legislative methodological assembly, and perhaps also a scientific/methodological "Thought-Police", to echo Imre Lakatos.[8] So, the reasoning adopted above leads us also into problems concerning the freedom of inquiry. Obviously, freedom and creativity are not unrelated. The degree of precision and explicitness we impose on the normative systems of science is connected with the way in which we choose, or are compelled, to institutionalize and organize inquiry.

Authoritarian regimes have tried to secure certain types of consensus by institutional measures: legislation, legal sanctions, indoctrination. This we find unacceptable, and we may doubt its long term success. It would seem little short of absurd, however, to secure creativity by similar means. We tend to believe that formal restrictions on a free and critical consensus formation and on a free and not too uncritical creativity are counter-productive in science, given the basic value of maximizing the expected epistemic utility of inquiry.

This may turn out to be an argument in favor of one aspect of the *status quo*. From the point of view of freedom, a network of separate and independent institutions (universities, institutes, departments) may, in fact, have its advantages. It means that codifications, bye-laws, charters and written rules are required and conceived within reasonably delimited and well defined social contexts.

The scientific community is no well defined social unit. Nor is it anything like a uniform profession.

There is one feature of the present scene, however, which is worth noting in this connection: the growing tendency to seek and to set up professional codes of ethics for distinct sub-groups within the scientific community.[9] But it is also notable that such codes concern, above all, the external norms and commitments of various scientific groups and professions, their relationship to society at large: social responsibility and accountability.

The issue of freedom of inquiry may be involved in the present topic in yet another way. The many-valuedness of our methodologies, the pluralism of ends and purposes in science, and the consequent possibility of value conflict – such as that exemplified by consensus and creativity – also make room for that kind of contingency which is a necessary condition of free and rational action. Where all follows of necessity there is little room for choice. So, this "open" logical structure of the normative systems of science also can be seen as contributing to making scientific inquiry a governable social process. Science policy is a name we give to one type of attempt to guide and govern this process, responsibly and accountably. In situations of contingent choice – where no alternative is forced upon us, causally or logically – we must *make* decisions. They can hardly be rational unless justified and supported by appeal to consistent normative systems.

Perhaps it is not compatible with (or consistent with) ideals of rationality to strive for an explicit and formal institutionalization of rationality *in general*.[10]

Consider for a moment the wider ideology of science comprising internal as well as external norms and values of inquiry. It is this ideology *in toto* which is necessary for a rational science policy. It is not only by upholding and defending the internal norms of science (methodological virtues, autonomy) that the scientific community can make its contribution. It cannot hope to justify itself in the eyes of a well educated public without appealing to values of public welfare. That is why one may wonder about the soundness of an attitude now on the wane in the scientific community: the view which idealized the "value neutral" scientist. This could perhaps now be described as the view that internal norms alone suffice to guide and justify *all* the actions of a professional inquirer. If this view is not actually inconsistent,[11] in a strict sense, it is certainly incoherent. There are arguments to show that even internal norms, notably those of methodology, are incomplete without a linkage with external norms. The point of contact is in the methodological requirement of *fruitfulness* or *relevance*. It is just not the case that all truths are equally fruitful and relevant, for theoretical or practical purposes.[12]

I am conscious of the problematic character of such reasoning, and not only because it may risk charges of triviality and rhetoric. Some of the most forbidding practical problems in connection with the "social responsibility" of science probably concern precisely its manner of implementation. Again, perhaps, the most realistic promises lie in the ideological and social organization of delimited sections of the scientific community.

Perhaps another way of stating one of the main ideas of the argument here attempted is this. Methodology has *two* basic values, because so does the overall ideology of science of which methodology is no merely contingent or arbitrary component. The two values could be called truth and relevance. Freedom from conflict in any one area of this ideology would be ensured only if the pursuit of (any) one of these two aims automatically secured the realization of the other by logical or some other kind of necessity. But there simply is no such benign necessity in science (or in philosophy, or in any other inquiry). Hence the need for rationality as something over and above either formal logical relations, or "natural" causal processes. Expressions such as "optimizing the truth-output", "maximizing expected epistemic utility", both rather artificial, signal the attempt to bring the two (truth and relevance) together, constructively.

Inconsistencies – logical contradictions – are not compatible with this notion of rationality. But there is nothing inherently irrational about conflict, and certainly not if the conflicting parties are values.

Humans may be rational or irrational in their ways of handling conflict.[13]

University of Oslo

NOTES

[1] 'The Foundations of Cognitive Activity: An Historical and Systematic Sketch', *Inquiry* 1976, pp. 131–50, and 'Norms of Inquiry, Methodologies as Normative Systems', printed in G. Ryle, (Ed.) *Contemporary Aspects of Philosophy* (Stocksfield: Oriel Press, 1977), pp. 1–13. Other references are given in these papers.

[2] The formulation is that used by Alonzo Church in Runes, *Dictionary of Philosophy*, article on "Consistency".

[3] This might be said to follow from the notion of coherence. Propositions may be incoherent (or uncoherent, non-coherent?) without contradicting each other.

[4] We do, indeed, find a great many *diverse* statements of the goals and aims of science, not all of them mutually compatible. My formulation ("to optimize the truth-output") is meant (is hoped) to capture a feature on which there ought to be agreement. The price I pay may be a considerable openness and vagueness, and a regrettably awkward form of words. My reason for choosing the word "optimize" rather than, say, "maximize" (or just plain "truth" and "knowledge") is, of course, that not all truths or all knowledge is equally valuable (fruitful, relevant). See below, Section 6, and Tranøy (1977), p. 9 f. - I would think that my expression, "to optimize the truth-output of inquiry" is very close to the decision-theoretic "maximizing epistemic utility", although I would not dare to say that they are in fact synonymous.

[5] Rawls, *A Theory of Justice*, p. 4.

[6] In Tranøy (1977) I have discussed some of the norms which may be applicable to the acceptance and rejection of propositions.

[7] Robert K. Merton's 'Science and technology in a democratic order' (1942) was republished in Merton, *Social Theory and Social Structure* (revised ed., Free Press, 1967, pp. 550–61) under the title 'Science and democratic social structure'. It is reprinted in Barry Barnes (ed.), *Sociology of Science* (Penguin Books, 1972, pp. 65–79), now under the title 'The Institutional Imperatives of Science'. Further references are found in Barnes.

[8] I have borrowed the term "Thought-Police" from 'Understanding Toulmin', a posthumous review by Imre Lakatos (and John Worrall) of Toulmin's *Human Understanding* (published in *Minerva* 14 (no. 1) pp. 126–143 (1976)). I am not certain, however, that the sense which Lakatos gives to the term, would also suit the purposes of the present paper.

[9] See for instance 'Codes of Ethics in the Social Sciences: Two Recent Surveys' by Paul Davidson Reynolds, in *Newsletter on Science, Technology and Human Values* (Harvard University), No. 18, January 1977, pp. 15–17.

[10] Perhaps what I have said here is also compatible with one of the main points in Feyerabend (1975)?

[11] The possibility of such inconsistency might be indicated thus: one can hardly, at the same time and consistently, believe that "science" is justified by its contributions to human welfare, *and* that the only professional obligation of the scientist is to the internal norms of science. To avoid inconsistency it would seem that he would have to stay out of science policy all together - which is not very probable in practice; it might not even be possible in theory for a professional scientist to avoid involvement in science policy.

A recent comment along similar lines on the notion of neutrality is Max Black, 'Scientific Neutrality. Between Truth and Irresponsibility', *Encounter*, August 1978, pp. 56–62.

[12] See also Tranøy 1977), p. 9 f.

[13] Attention might also be called to Dorothy Nelkin, 'Changing Images of Science: New Pressures on Old Stereotypes', *Newsletter 14 of the Program on Public Conceptions of Science*, January 1976 (= *Newsletter etc.* as quoted in note 9 above), pp. 21–31, especially pp. 26 ff, "Internal Contradictions". "All social systems contain contradictions; those within science often derive from ambiguity regarding its cognitive and pragmatic dimensions" (p. 26). It seems, however, that Nelkin does not intend "contradiction" in a strict logical sense, but rather in the sense of "conflict" as I have used this word above. Obviously, we are here touching on the problem of the nature of non-formal contrariety or opposition, which is probably also relevant to my topic.

REFERENCES

Black, Max: 1977, 'Scientific Neutrality. Between Truth and Irresponsibility', *Encounter*, August 1978, pp. 56–62.

Church, A.: 1965, 'Consistency', in Runes, D. *Dictionary of Philosophy* (15th Ed.), Littlefield, Adams and Co., Totowa, N.J., p. 65.

Feyerabend, P.: 1975, *Against Method*, NLB, London.

Lakatos, I. and John Worrall: "Understanding Toulmin" in *Minerva* **14** (no. 1), pp. 126–143 (1976).

Merton, R. K.: 1942, 'Science and technology in a democratic order' under the title 'Science and democratic social structure' in Merton, R. K., *Social Theory and Social Structure*, Free Press, 1967, and 'The Institutional Imperatives of Science' in Barnes, B. (ed.), *Sociology of Science*, Penguin Books, (1972).

Nelkin, Dorothy: 1976, 'Changing Images of Science: New Pressures on Old Stereotypes', *Newsletter of the Program on Public Conceptions of Science*, (= *Newsletter* etc., as given in note 9 above), No. 14, January 1976, pp. 21–31.

Reynolds, P. D.: 1977, 'Codes of Ethics in the Social Sciences: Two Recent Surveys', in *Newsletter on Science, Technology, and Human Values*, No. 18, Harvard University, Aiken Computation Laboratory 231.

Tranøy, K. E.: 1976, 'The Foundations of Cognitive Activity: An Historical and Systematic Sketch', *Inquiry* **19**, pp. 131–50.

Tranøy, K. E.: 1977, 'Norms of Inquiry, Methodologies as Normative Systems', in G. Ryle, (ed.), *Contemporary Aspects of Philosophy*, Oriel Press, Stocksfield, 1977.

JULES VUILLEMIN

THE INFLUENCE OF REASON
ON THE ORIGIN OF SCIENCE

When we speak of rationalism and examine the influence of reason on the origin of science, we must be careful that our concept be neither too narrow nor too large.

Our concept would be too narrow if limited to one of its historical forms. It would become too large, were we to count as a rationalist everybody who trimmed with empiricist garments some idea borrowed from rationalism.

The two following conditions are necessary to define rationalism:

(1)   suprasensible entities are to be accepted as irreducible principles,
(2)   these principles can be the object of a pure intellectual knowledge.

These criteria suffice to separate reason from empirical knowledge and opinion. They do not suffice, however, to account for a possible influence of reason on the origin of science. Parmenides himself, after having distinguished reality from appearance, brought great care to give us, in the second part of his poem, some hints for deriving from the rational reality the host of sensible appearances. We must therefore add a third criterion, namely

(3)   that the suprasensible entities be able to save the appearances.

Not unseasonably, the word *ratio* means reason and cause.

If these three criteria are enough it would be too much to require e.g. with the Eleatics that the suprasensible entity be unique, or with most rationalists until Descartes that the suprasensible organizes the world into a linear hierarchy of perfections. Monism and linearity in the scale of perfections have been extremely important in the history of ideas. But to adopt them as criteria of rationalism would amount to putting Plato's ideas or Descartes dualism outside rationalism.

One can rightly wonder that such a meagre and shadowy rationalism played a decisive role in the history of science. Instead of dwelling on idle speculations, let us try to show the how and why of an unexpected

R. Hilpinen (Ed.), Rationality in Science. 203–208.

fecundity on three examples concerning the origins of Greek mathematics, of Greek astronomy and of the theory of functions.

I

What makes pythagorean rationalism pure and specific, i.e. immune from any empirical contamination, is that the appearances are saved by using only rational principles. Every explanation in astronomy, in harmony or in geometry must therefore rely on a unique and universal principle satisfying our three criteria. The reason being conceived of as the faculty of ratios, it is agreed that everything is number. Natural numbers and their ratios exhaust the whole universe.

Music, that of the lyre and that of the spheres, entered as well as the arithmo-geometry into the numerical scheme. No wonder that the school was shocked when Pythagoras discovered that there is no ratio between the diagonal and the side of the square. Theorems belonging to the well established theory of proportions such as the possibility of interverting in a proportion the extreme or the mean terms became invalid. Being foreign to reason, the sensible field of space, the $\chi\omega\rho\alpha$ made of a mixture of being and not-being could afford no help. Useless too was a composite system of knowledge such as Aristotle's, exploiting a multiplicity of categories and even splitting quantity into discrete numbers and continuous magnitudes, each of which having its own incommunicable laws. Rationalist standards are more ascetic or more exacting. New methods had then to be invented by seeking for more general ideas of reason.

The celebrated fifth book of Euclid, attributed to Eudoxus and where an axiomatic and contextual definition of proportion or rather of equal proportions was given, did not follow out the spirit of rationalism.[1] As Aristotle says, Eudoxus made sensible the ideas; renouncing the clear and isolated evidence of the rational principles, he relied on an axiomatic system, the truth of which is known only through experiencing its consequences. In this elegant and powerful system, there is a pragmatic element alien to the rationalist requirements.

Plato's dialogues remind us of the narrow solution which was open to the rationalists. They had to widen the concept of number by accepting infinite definitions, i.e. infinite developments of a ratio according to a determinate rule. Theorodos constructed such definitions, the successive approximations of continued fractions for the square roots of the numbers up to 17. Theaetetus generalized the notion of a root and, as Plato says, included all "the innumerable roots under one name or class".

The narrowness of a rationalist program, not far from certain constructivist conceptions, precluded for the time being every possible answer to the paradoxes which Zeno of Elea and Diodorus of Megara had developed concerning the continuum. We do not know very much about what has been proposed by the platonic school. It is possible and even probable that, after Theaetetus, the main trends in Greek mathematics led away from rationalism. These foreign developments however had been possible only because of two characteristic advances both due to rationalism. First the field of science and of its methods were defined within very narrow bounds: only finite constructions with natural numbers were allowed. Secondly, this united field had to split up, because the accepted methods did not fit in with their object. It may seem paradoxical to consider a crisis as an advance. But there are two component parts in rationalism: ideas and methods, principles and proofs. Now, while empiricism can simulate the successes of rationalism, there is no counterpart of a *reductio ad absurdum* in the field of senses. By stumbling against experience, reason is reflected on itself and thus acquainted with its highest principle, according to which the world must be self consistent. Finally, by admitting infinite definitions, Theaetetus showed how to remain faithful to reason while leaving off believing that the development of a fraction has to be finite in order to be recognized as rational. A first step had been taken towards the conquest of the continuum.

II

We are told by Plato that the same Pythagoreans were the first to formulate the astronomical problem: to save the heavenly appearances by a superposition of circular and uniform motions. Why did they regard a circular and uniform motion as rational ? Their argument may tentatively be reconstructed as follows:

(1)   The world is finite (as is every number). Like an infinitely great number, an infinitely great world would be undeterminate.

(2)   The diameter of the world is therefore finite.

(3)   A rectilinear motion can neither be continuous nor keep the same direction. When reaching the extremity of the world,

        such a motion could only continue by reversing its direction, which introduces a discontinuity and a contrariety.

(4)   Such are the sublunary motions, characteristic of beings which are neither always moved nor always resting, but participate to the disorders of generation and corruption.

(5)   If there are perfect, i.e. perpetual motions, they must be continuous and keep their direction at the same speed. They must therefore be circular and uniform, as it happens with the heavenly motions.

This scheme was as narrow to save the astronomical appearances as the scheme of number was for saving the geometrical appearances. Only the theories of the homocentric spheres or at most the theories with movable epicycles could comply with the rationalist requirements. A system like Ptolemaeus' could not, since with the excentrics was recognized that the circular motions are not uniform. Here too, the jacket had been ill tailored. Some among its consequences show how unfortunate for the development of astronomy had been the prejudice of reason; the discovery of positive laws had to be delayed until Kepler; the representation of inertia was prevented; the composition of the circular uniform motion was ignored; any mathematical identity between a pendular motion, a circular motion and its projection on the diameter as an harmonic motion was precluded, although Ptolemaeus had established the last correspondence but only to justify the existence of an harmonic motion by reducing it to a circular one and without seeing that they were functionally identical.[2]

Meanwhile, a false hypothesis is better than no hypothesis on the condition that it is narrow enough to be easily falsified and such was the astronomical idea of the Greek rationalism.

III

The last example will be taken from a more modern case, that of the theory of functions as initiated by Descartes.[3]

What is required from a function to be an idea of reason? Cartesian mathematics are purely intellectual; they differ from Euclid's geometry in as much as the spatial imagination of the figures is completely subordinated to the understanding. In order to be rational i.e. to be the object of a clear and distinct idea, the extension does not admit any sensible reality except the capacity of representing algebraic equations. Only the finite operations of algebra, which are completely distinct for our understanding, are therefore considered as rational. On the contrary, Descartes rejects every expression which requires an infinite number of algebraic operations, as do trigonometry and calculus. Far from achieving the leibnizian program of a complete analytical geometry, he severs the algebraic curves from the mechanical ones. He knows how to construct mechanically the logarithmic curve and the logarithmic spiral but, as the motions he needs for his construction are incommensurable, he considers that they do not belong to the proper field of the Analysis.

A mechanical curve has indeed something indistinct or infinite which bypasses our power of understanding. It may be used as an heuristic tool; it lacks a proper methodical foundation.

The history of the theory of functions is nothing else than the progressive widening of Descartes' conception. Perhaps the most important occasion for this widening was offered by the old pythagorean problem of the plucked strings, when Fourier solved the problem "which consists in developing any function whatever in an infinite series of sines or cosines of multiple arcs" and extended "the same results to any function, even to those which are discontinuous and entirely arbitrary".[4] The story repeated itself: a narrow concept framed according to a rational criterion, a case of experience clearly adverse to the scheme, the elaboration of a new scheme. Now this elaboration obeyed pragmatic rules which had nothing to do with the ideal of rationalism, now the reconstruction proceeded according to an organic renewal of this ideal. That was for instance the case, when, challenged by the growing species of teratological functions, Jacobi and Felix Klein tried to define what properties required a function which could be considered as a reasonable one (*die vernünftige Funktionen*).[5]

It has been and it is fashionable to despise reason. A stubborn ignorance of history encourages in believing that it has been and it is a barren faculty, at the best a faculty for tautologies.

But even tautologies may happen to bear the unexpected. And it was not by mere chance that Anaxagoras, who may be considered according to our three criteria as the first rationalist philosopher, found something questionable and unexpected in the most abstract principle of reason, namely in the excluded third when applied to infinite sets.

Collège de France

NOTES

[1] Vuillemin, J., 1977, 'Définition et raison: le paradigme des mathématiques grecques', *Actes du Congrès de la société grecque humaniste Athènes-Pélion 1975*, Athènes, pp. 273–282.

[2] Ptolémée Cl., *Syntaxe mathématique*, Livre XIII, ch. II, éd. Heiberg, Γ΄, β΄, pars II, pp. 529–534; Duhem, P., *Le système du monde*, t. II, Hermann, Paris, 1914, pp. 234–237.

[3] Vuillemin, J., *Mathématiques et métaphysique chez Descartes*, Paris, P.U.F., 1960.

[4] Oeuvres de Fourier, éd. Darboux G., t. I. (*Théorie analytique de la chaleur*), Gauthier-Villars, Paris, 1888, p. 224.

[5] Klein, F., *Elementarmathematik vom höheren Standpunkte aus*, III, Erste Aufl., Nachdruck 1968, Springer, Berlin, 1968, p. 50; p. 128.

PAUL WEINGARTNER

NORMATIVE CHARACTERISTICS
OF SCIENTIFIC ACTIVITY

This paper is concerned with the problems described in the research project "Foundations of Science and Ethics" as follows:

The most general aim guiding the scientific enterprise is to obtain non-trivial truth and theoretical understanding. We shall outline in a preliminary way some conjectures concerning the normative features of scientific inquiry. Some norms are general in nature and estend through the various sciences as global restrictions. These include the already mentioned concern to seek truth and increase theoretical understanding. Interest in simplicity, consistency, coherence, and comprehensiveness are other general and perhaps invariant goals of contemporary science. One fundamental precept of science is to base conclusions on all empirical information that is available in the scientific community.

The paper is divided into four sections: The first section deals with the most fundamental principles in the sense of most general goals of all sciences and of all scientific research work. The second section is devoted to principles of general methodology which are common to all methodologies of the different scientific disciplines. The third section is concerned with principles which occur in the different methodologies of the different scientific disciplines. The fourth section is concerned with principles in normative disciplines.

1. FIRST LEADING PRINCIPLES

1.1 *Truth*: Is truth the aim of science?

It seems not. For:

1.1.1. (1) If the aim of science is to find satisfactory explanations then the aim of science cannot be to know the truth, since the *explicantia* (the explanatory statements) are usually not known to be true. Moreover something not known could hardly be an aim; thus truths not known to scientists could hardly be an aim for them. (2) But as Popper says: "It is the aim of science to find satisfactory explanations".[1] (3) Therefore truth seems not to be the aim of science.

1.1.2. (1) A science is a system of statements referring to a certain

209

R. Hilpinen (Ed.), Rationality in Science. 209–230.

structure of objects. (2) But a system of statements does not have an aim. (3) Therefore truth cannot be the aim of science.

1.1.3. (1) If something is an aim or end or goal it must be sufficient, not only necessary like the means. (2) But truth is not sufficient for science. This is seen as follows: (a) A discipline which has established all particular truths formulated in singular statements in a special field "possesses" in some sense the truth in that field. But sciences search for theories as general as possible to explain the particular with the help of the universal. (b) Though $p \rightarrow p$ and $x = x$ are universal truths they are not aimed by sciences since these truths are not interesting, they are lacking content. Thus from (a) and (b) it follows that truth is not sufficient for science. (3) Therefore truth seems not to be the aim of science.

1.1.4. On the other hand it seems correct to say that the aim of science is truth. For: (1) The task of science is to find out what is (the facts about the world and the universe). (2) But instead of saying that something is a fact or something *is so* we may say that a sentence, representing it, is true: "What do you mean by there being such a thing as Truth? You mean that something is SO"[2]. (3) Therefore the task of science is to find out the true sentences (representing facts), or "to find out the truth".

1.1.5. *Proposed Answer*

1.1.51. It seems to me to be a fact what Aristotle says in the first statement of his metaphysics; "All men by nature desire to know . . ."[3] One may even weaken this claim in a twofold way to a minimal principle about mankind: One may replace 'all' by 'almost all' or 'statistically all' to allow some few extravagant exceptions caused by exceptional conditions. And one may replace 'desire to know' by 'desire to know more or better relative to that what and how they know at the present time and relative to their interests and abilities." The minimal principle can then be stated thus:

Almost all men (by nature) desire to increase their knowledge relative to their interests and abilities.

1.1.52. In the following I want to propose a sufficient condition of what I would call a common goal (end, aim) for mankind:

1.1.521. X is a common goal for mankind if all men desire X.

Also here it seems to me to be possible to weaken the antecedent (and therefore to strengthen the thesis):

1.1.522. X is a common goal (end, aim) for mankind if almost all (statistically "all" or highly significantly all) men desire X.

On the other hand, if all men of some social groups or society, but no other people outside the group or society, desire something it would be a sufficient reason to speak of a subjective goal.[4] According to 1.1.521 and 1.1.522 one may say that knowledge or the increase of knowledge relative to interest and ability is a common goal (end, aim) for mankind.

1.1.53. A very important type of knowledge is scientific knowledge. Thus all scientific knowledge is knowledge, but not vice versa. Scientific knowledge is a sufficient (even if not necessary) condition or *one* possible way to reach knowledge and this means to satisfy a goal of mankind. If Z is a goal (as knowledge is a goal of mankind according to 1.1.52) and Z_1 is a sufficient condition for getting Z (a possible way to reach Z) then one may call Z_1 a subordinated goal in respect to Z (or a sufficient means in respect to Z). In this sense one may say that scientific knowledge is a subordinated goal of mankind.

1.1.54. Testable or confirmable truth (or approximate truth) is a necessary condition for scientific knowledge. For scientific knowledge one has to require the thesis: If the person a knows scientifically that p, then p is testable or confirmable true (or approximate true). In other words: For getting an adequate concept of scientific knowledge one would not allow to say that a knows that p if in fact p is neither testable nor confirmable (for a or at least for a scientific community who can inform a).[5] In addition one has to realize that the actual situation in scientific research is often such that we cannot have testable or confirmable truth as a necessary condition for scientific knowledge but only testable or confirmable approximate truth.[6]

A further necessary condition is that the truth which is approached by the scientific disciplines is informative and contentful. The search in science is not for mere tautologies or for uninteresting singular truth but for comprehensive truth expressed by law statements. Thus scientific knowledge requires informative truth (or approximate truth) which is testable or confirmable. But we may say that informative, contentful truth (or approximate truth) which is testable or confirmable is a necessary and sufficient condition for scientific knowledge. And therefore such a truth (or approximate truth) is also a subordinated (cf. 1.1.53) goal in respect to knowledge in general (which is a goal of mankind).

1.1.55. From what has been said so far it follows that though scientific knowledge and informative and testable truth (and approximate truth) are subordinated goals in respect to knowledge as one of the goals of mankind, they are nevertheless the highest goals of scientific activity.

1.1.56. It is a historical fact and it has been a historical experience of mankind that informative, testable and confirmable truths were the result of scientific activity; and in most cases: of scientific activity which was done by a scientific community who acted according to methodological rules. Today, in the twentieth century, scientific activity and scientific approach shared by a community of scientists and governed by methodological rules is a nècessary condition for reaching informative, testable truth (or approximate truth) and scientific knowledge.

1.1.57. In order that scientific activity proceeds more efficiently to reach the goals of truth and approximate truth it has to be ruled by rules and norms of methodology. Or to put it into a true conditional: If scientific activity is not ruled by methodological rules, it does not lead efficiently to informative and testable truth and approximate truth. These rules and norms are also subject to scientific test, criticism and confirmation in order to be revised and improved permanently. A test for the validity of such a rule or norm consists mainly in an empirically testable *modus tollens* argument of the following form: if the scientific activity or research-activity does not proceed according to this or that rule, it either does not lead to truth or approximate truth (for instance, it leads to false statements), or it does not reach this goal efficiently. In this sense every scientific discipline has also the task of establishing and testing the methodological norms which are specific for that discipline.

1.1.58. Because of what has been said in 1.1.56 and since rules and norms are not true (false) and not approximately true but valid (invalid) or approximately valid, one has to add validity of norms if speaking more completely of the goal of scientific activity: the goal (end, aim) of scientific activity is testable or confirmable, informative truth and validity or approximate truth and validity.

1.1.6. Commentary to the objections:

1.1.61. (ad 1.1.1): (1) Though the *explicans* is usually not known to be true, we are not totally ignorant about it: it is confirmed and corroborated if it withstands severe tests and explains already known facts. By development of this process the explicans can be known to be approximately true. (2) "Something not known" is ambiguous, it can mean not known specifically, but known in general: before Einstein invented his theory of relativity this theory was not known ("specifically") and thus it couldn't have been an aim or goal; however it was known (by some "in general") that there is a need for a more informative theory of space and time which is a better approximation to the truth than Newton's theory, and a theory with such characteristics could be an aim or goal for a scientific approach.

1.1.62. (ad 1.1.2): The expression 'science' is ambiguous: (1) In one sense it means a system of statements referring to a certain structure of objects. And in this sense it is correct (as said in the objection) that such a system does not have a goal or aim; on the contrary – if the system of statements is tested and confirmed and includes some laws or theories of highly explanatory power then we may rather say that the system itself *is* an aim or a goal. (2) In another sense 'science' means an activity, the scientific activity, the research-activity, the research-work. And in this sense one can say correctly that the scientific activity has an aim or goal; to find a system of informative hypotheses and laws put together in a theory which is true or a good approximation to the truth.

1.1.63 (ad 1.1.3): The objection is correct in the sense that it is more accurate to say that *informative* and *comprehensive* truth (or approximate truth) and validity (or approximate validity) is the aim of science instead of just saying that "truth" is. It should be mentioned that what is informative and what is comprehensive is relative to the different scientific disciplines: For instance, $x = x$ is an irrelevant and a trivial truth in most disciplines but it may be a very interesting truth in logic (identity theory)[7].

1.2. *Consistency:* Is consistency a goal of science?

It seems not. For:

1.2.1. (1) Every true (or valid) sentence and every set of true (or valid) sentences is consistent. (2) If testable or confirmable informative truth is the goal of science then consistency is included. (3) Therefore consistency seems not to be a goal of science.

1.2.2. (1) If consistency is a goal of science the principle of non-contradiction must be generally acceptable among scientists. (2) But as Putnam says, the principle of non-contradicion is not generally acceptable. What is acceptable according to Putnam is a "Minimal principle of Contradiction" which is much weaker than the principle of non-contradiction: "Not every statement is both true and false."[8] (3) Therefore consistency seems not to be a goal of science.

1.2.3. (1) If there are strong logical reasons that something, A, cannot be achieved, then A cannot be a goal of science. (2) But there are strong logical reasons that consistency cannot be achieved for number-theory, analysis and set-theory: "Such a consistency proof, however, can be regarded as *absolute* only if the metatheory is unimpeachable. Otherwise, it carries conviction only *relative* to the degree in which we are convinced of the consistency of the metatheory. For certain theories, such as number theory, analysis, and set theory, it looks hopeless to find a suitable metatheory that would not be at least as suspect as these theories themselves . . ."[9] Since almost all sciences presuppose and use parts of number theory, analysis and set-theory, consistency cannot be achieved for them either. (3) Therefore consistency seems not to be a goal of science.

1.2.4. Contrary to the objections 1.2.1–1.2.3 Hilbert says that the principal task of the axiomatic method is to be able to make sure that inconsistencies cannot occur at all because of the axiomatic construction: ". . . die prinzipielle Forderung der Axiomenlehre muß vielmehr weitergehen, nämlich dahin, zu erkennen, daß jedesmal innerhalb eines Wissensgebietes auf Grund des aufgestellten Axiomensystems Widersprüche überhaupt unmöglich sind."[10]

1.2.5. *Proposed Answer.*

If A is a goal and if B is a necessary condition for A then B is a necessary condition for reaching goal A, i.e., B is a means for A (cf. 1.1.53). Now, consistency is a necessary condition for truth and validity and thus testable and confirmable consistency is a necessary condition for testable and confirmable truth and validity. Therefore consistency and testable and confirmable consistency are means for the general goal of science which is testable and confirmable truth and validity. But they are also means for the weaker goal of approximate truth and validity. This is so because approximate truth and validity of a theory implies — though it allows some false (invalid) consequences of the theory – that these consequences have to be factual, i.e. not logically false (not inconsistent). Thus consistency is a means for the goal of science.

1.2.6. Commentary to the objections:

1.2.61. (ad 1.2.1): It is correct – as said in the objection – that consistency is included in truth and validity; therefore it was said in 1.2.5 that consistency is a *means* to the goal (end) of science, not the very end itself. However, since the stronger result (ultimate end) is in several cases unattainable, it may be a great thing to reach even the weaker result, i.e. consistency. This is also important in those cases where we can have at least *approximate* truth or validity.

1.2.62 (ad 1.2.2): The first thing to be observed here is that the expression "the principle of non-contradiction" is not unique. Rather there are several versions of the principle of non-contradiction which differ in strength. For instance the following four principles are different in strength:

1.2.621. The thesis '$(p) \sim (p \wedge \sim p)$' is logically true, whereas p can have one of the values 'true' or 'false'.

1.2.622. The thesis '$(p) \sim (p \wedge \sim p)$' is logically true.

1.2.623. At least one member of the pair p, $\sim p$ is false.

1.2.624. At most one member of the pair p, $\sim p$ can be true (or: can have a designated value).[11]

The first (1.2.621) is most restrictive (very strong) and allows only classical two-valued logic. The second (1.2.622) is weaker; it is valid in several systems of many-valued logic but fails for instance in the three-valued systems of Łukasiewicz (with the value 'T' alone designated), Post, Bochvar and Kleene. The third principle (1.2.623) behaves similarly; it is true in several many-valued systems but fails in the system of Łukasiewicz, Bochvar and Kleene. The fourth (1.2.624) is the most tolerant principle and is valid in all two and many-valued systems of logic known so far. It seems to be the weakest version of the principle of non-contradiction.

1.2.625. I think Putnam is right that the principle '$\sim (p \wedge \sim p)$' is not generally acceptable if one of the versions 1.2.621, 1.2.622 or 1.2.623 is its interpretation; moreover he seems to be right if any version which is stronger than 1.2.624 is taken to be the principle. On the other hand I do not see any reason why the most tolerant version, i.e. 1.2.624, should not be generally acceptable for all sciences.[12]

1.2.63. (ad 1.2.3): It is correct that there are no absolute consistency proofs for the most general theories of mathematics for instance for set theory. But although this is correct and although the second premiss – i.e. saying that almost all sciences use parts of number theory, of analysis and of set theory – is correct too the conclusion that consistency cannot be achieved for them is too strong and does not follow: The reason is that those parts of the general mathematical theories in question which are used by the empirical sciences can – at least in principle – be either treated within the finite domain or within the denumerable domain. And this means that those theses (axioms) which are viewed as to be most problematic (continuum hypothesis, axiom of choice, strong axioms of infinity, axiom of replacement, unrestricted axiom of power set) concerning consistency are not presupposed. However, the situation is different (from set theory) for number theory and analysis. There are consistency proofs with constructive methods for number theory (Gentzen, Gödel). Though there is no such proof for the whole classical analysis there are consistency proofs with constructive methods for a part of it the so-called constructive analysis and even for a further part which is restricted to predicative functions (predicative analysis). But there are strong reasons that the theories in physics and in other empirical sciences do not need stronger means than those provided by number theory, constructive and predicative analysis. Thus it is not impossible to obtain consistency results in those sciences.

1.2.64. (ad 1.2.4): The so-called 'Hilbert Program' consisted essentially of two parts. First in formalizing all or great parts of mathematics by constructing a formal system with the help of the axiomatic method. Examples at that time were *Principia Mathematica*, Peanos' axiom system for number theory or Zermelo's axiomatization of set theory. The second step was to show – as the passage quoted in 1.2.4 says – that the process of deriving theorems in a formalized (axiomatic) system, just by applying the rules of inference to the axioms can never lead to any contradiction, i.e. to a derived end-formula of the form $1 = 2$. The arrangements by which this step (i.e. the claim that the derivation of a contradiction is impossible) was to be shown (formulated within a metatheory, which Hilbert called "metamathematics" or "proof theory") had to be of very elementary ("finitary") character, i.e. not allowing steps which intuitionists or other mathematical schools found objectionable like the tertium non datur for infinite sets, impredicative concepts, indirect existence proofs etc.[13]

The theorems of Gödel (1931) showed that the original Hilbert Program (as roughly outlined above) cannot be carried through.[14] Therefore the way proposed in 1.2.4 to obtain consistency results is not possible. As a result sciences are left with finding out relative consistency proofs and with getting new consistency results by applying restrictive conditions (cf. 1.2.63).

1.3. Summary of sections 1.1 and 1.2

First it was assumed that almost all men by nature desire to increase their knowledge relative to their interests and abilities (1.1.51). And this was called a common goal (end, aim) for mankind (1.1.52). Then it was argued that scientific knowledge is a very important type of knowledge and can therefore be called a subordinated goal of mankind (1.1.53). For scientific knowledge (in the sense of a system of statements, not in other senses for example as a dispositional property of men) a necessary and sufficient condition is informative contentful truth (or approximate truth) which is testable or confirmable (1.1.54). In the following chapters 1.1.55 and 1.1.56) it is argued that scientific knowledge presupposes scientific research (approach) which again presupposes general and specific methodological principles (as necessary conditions). In 1.1.57 rules and norms of methodology are added under the informative testable sentences since they also are criticized and confirmed during the scientific procedure. Speaking more completely of the goal of scientific activity one may say then that testable informative truth (approximate truth) and validity (approximate validity) is the goal of scientific activity.

In 1.2 it is argued that the search for consistency is presupposed by the search for truth and validity such that consistency can be called a subordinate goal in respect to the main goal of scientific activity (testable informative (approximate) truth and validity).

1.4. First normative principles

Every goal represents the highest value (or the most preferable goal) relative to its subordinate goals and to its means. Goals (representing such values) may be reformulated by rules or norms. Or one may derive rules or norms from goals by presupposing the following premiss: If something A is a goal of (for) some human activity B, then it (A) should be searched for (by B). Specifically: If testable informative truth and validity are the goals of scientific activity then it should be searched for and approached by scientific activity. The same holds for consistency. Therefore the first normative principles of scientific activity can be formulated thus: Search for testable informative truth and validity (approximate truth and approximate validity), including consistency.

2. NORMATIVE PRINCIPLES OF GENERAL METHODOLOGY

2.1. Normative principles of general methodology, level I.

A sentence n is a *first-level* normative principle of general methodology if it satisfies the following conditions:

2.1.1. *n* is a general (universal) norm (normative sentence) or rule applied in all or in most research activities of the different scientific disciplines.

2.1.2. *n* is testable at least in the following way: if *n* is violated then one of the first normative principles (cf. 1.4) is violated (at least to some degree) in one of the following two senses. (a) if *n* is violated then there is serious danger that false or invalid (even perhaps inconsistent) statements and norms may enter. The phrase "there is serious danger" means that it has occurred so (that false and invalid statements enter when *n* is violated) in (statistically) most cases in the historical and recent development of the sciences. Therefore there is significant evidence for the hypothesis that it will occur again. (b) if *n* is violated then the requirement for *contentful* and *informative* truth (validity) or approximate truth (validity) is violated too.

2.2. Examples of normative principles of general methodology, level I.

2.2.1. Base your hypothesis (thesis, theory) on all scientific information available.

If this rule is violated hypotheses may be based on interpolations which are too risky; or important new information which would allow severe tests may be ignored.
Caution: There are situations where all the information available in a certain field can hardly be interpreted consistently.

2.2.2. Make your concepts as precise as possible.

If this principle is given up one commits equivocation which leads to invalid inferences and hence to false conclusions (thus violating the rule of searching for truth).
Caution: Precision and exactness is not an end but a means. Therefore empty preciseness should be avoided. In many situations introducing distinctions (between two or more meanings) is necessary to achieve more preciseness.

2.2.3. Give your thoughts (more specifically: your hypothesis, your theory) a logical and/or mathematical structure.

A violation of this rule makes it more risky to commit an error; since without a logical and/or mathematical structure internal consistency and external consistency (i.e. compatibility with experimental or other basic results or with other hypotheses or theories) is hardly testable.

Caution: Avoid empty symbolism and empty formalism; do not confuse formalization with abbreviation.

2.2.4. Try to confirm your hypothesis by seriously testing and criticizing them. Look for interesting exceptions when testing your hypothesis.

If this rule is neglected a hypotheses may get a huge quantity of confirming instances, but nevertheless be false; the reason for the many positive instances is that it was tested within a scope where it was clear from the beginning that it must come out true.

Caution: There is a difficult problem about the question what are *severe* and *serious* tests.

2.2.5. New theories (hypotheses) should include the correct results of the old theories (forerunners) as special cases. If 2.2.5 is violated, the new theory will not satisfy some correct results and will therefore have some false consequences.

Caution: Since the framework and concepts may change from an older theory to a new one it is inaccurate to speak of "special cases". For example, mass (in the sense of classical physics) is scarcely a special case of relativistic mass (in Einstein's theory).

2.2.6. If a hypothesis or theory has been criticized or refuted as a general theory then look for partial sections (domains) in which it is still correct.

If this rule is violated one does not know enough of old hypotheses (theories) in order to create new ones so that there is real danger to invent a new hypothesis (theory) which has false consequences in a domain where the old one was correct. Rule 2.2.6 is closely connected with rule 2.2.5.

Caution: Do not use this rule to save a hypothesis or theory by all possible means.

2.2.7. Do not confuse conceptual objects with ontological (empirical) ones. And as a consequence: Do not construct (introduce) non-conceptual entities when there are only abstract conceptual terms.

This rule is one form of Ockham's razor. If violated, false assumptions about objects of the discipline in question enter into the system.

Caution: There are cases where it is necessary to invent (construct) new entities, especially when *ad hoc* hypothesis are necessary.

The following two rules of Newton are concerned with more specific cases of the rule 2.2.7:

2.2.8. "We are to admit no more causes of natural things, than such as are both true and sufficient to explain their appearances."[15] A violation would mean that the causes (or in the explanation: one of the premisses) are not true or that they are not sufficient. In both cases the explanation is not correct, either containing false premisses or not sufficient ones to derive the conclusion. Thus again a violation of this rule allows false statements to enter.

Caution: In many cases scientists do not know whether their hypotheses (theories) are true. But they can have methods to find out when one hypothesis has a better approximation to the truth than another.

2.2.9. "Therefore to the same natural effects we must, as far as possible, assign the same causes."[16] One may rewrite this rule in the following way: To the same phenomena apply the same explanations. A violation would mean to introduce new causes (premisses, explanations) which are stronger than the sufficient ones and such a policy increases the risk of false or at least superfluous assumptions.

Caution: There are cases where stronger assumptions (in an explanation) lead to interesting new predictions and thus to new severe tests of the hypothesis (or theory) in question, i.e.: Don't forget to look for better explanations.

The rules (2.2.1–2.2.9) mentioned so far are connected with the goal to approach scientific knowledge – i.e. with the goal to reach informative and testable truth and validity (or approximate truth and validity) – in the following way: if one of these rules is violated then the above goal is violated too, in the sense that there is a risk that false statements enter into the scientific system (cf. 2.1.2 (a)).

The rules 2.2.10 and 2.2.11 of general methodology are such that a violation of them means a hindrance to achieve contentful and informative truth and validity (approximate truth and validity) (cf. 2.1.2 (b)).

2.2.10. Explain the particular with the help of the universal.

If this rule is violated the particular is explained again with the particular; i.e. there are no universal statements in the explanation and consequently no law-statements. This again means that the explanans does not contain enough contentful and informative truth, which violates the goal described in 1.1.54.

Caution: Don't identify universality with strict universality. There are many fields in which one can reach at best probabilistic or statistical laws (as in sociology and history but also in thermodynamics).

2.2.11. Explain the concrete with the help of the abstract.

A violation of this rule leads – analogous to a violation of the rule 2.2.6 – to the acceptance of an explanans which lacks content and information i.e., it violates the goal described in 1.1.54.

Caution: Searching for abstract laws does not mean rejecting application and concrete tests, which are important and necessary for getting scientific knowledge.

The rules 2.2.10 and 2.2.11 have been proposed already by Greek Philosophy: Explanation of the visible, particular, concrete world by invisible, universal, and abstract principles.

2.3. Normative principles of general methodology, level II.

A sentence n is a *second-level* normative principle of general methodology if it satisfies the following conditions:

2.3.1. n is a general (universal) norm (normative sentence) or rule applied in all or in most research activities of the different scientific disciplines.

2.3.2. n is testable at least in the following way: if n is violated then the scientific activity is prevented from proceeding more efficiently to the goal required by the first normative principles (cf. 1.4); i.e. to the goal of informative and testable truth (approximate truth) and validity (approximate validity) and to the goal of increasing our scientific knowledge.

2.4. Examples of normative principles of general methodology, level II.

2.4.1. Try to create hypotheses which make new predictions (in the widest sense of the word)[17] and which make suggestions for new kinds of tests.

A violation of this rule allows hypotheses not so efficient for the growth of scientific knowledge. Caution: Avoid to ask for such new predictions or tests which trivially satisfy the original hypothesis.

2.4.2. Take the scientific tradition into account.

Although the knowledge of the scientific tradition is not a necessary condition to find the truth it nevertheless is a guidance of considerable importance in the following sense: (a) we learn from the mistakes of our forerunners, (b) we learn the correct results of our forerunners in order to incorporate them into the new hypotheses (cf. 2.2.5).

Caution: Don't stick to the tradition; a reinterpretation of tradition is not a scientific approach outside the history of science or the history of philosophy.

2.4.3. Keep away from subjectivism: Do not allow subjective terms to enter into the assumptions, definitions and theorems of your theory.

Subjective terms weaken the results and make them hardly testable ("it seems to me that *A* is *B*" is weaker than "*A* is *B*" and practically unfalsifiable). Therefore a violation of this rule leads to a hindrance of the growth of knowledge. Caution: When searching for objectivity don't confuse objectivity with non-applicability.

2.4.4. Make use of the hypothetico-deductive method.

This method had been proved to be most efficient for scientific approach. A violation slows down the progress of scientific knowledge. Caution: There are sometimes situations where other methods may be more appropriate.

2.4.5. Look for simplicity if compatible with truth (validity) and other necessary requirements.

A violation would not necessarily lead to false assumptions but it would prevent you from proceeding efficiently to the goal of informative and testable truth and validity (approximate truth and validity). Caution: Simplicity is a means, not an end; unification is (besides truth and validity) much more important.

2.4.6. Look for systemicity, i.e. for interrelated laws and principles.

A violation prevents efficiency in drawing consequences and in testing. Thus it violates also 2.4.4. This makes it more difficult to approach testable truth and validity. Caution: Also systemicity is a means and not an end. One should never take the risk of allowing false statements in order to have systemicity (for instance by proposing very strong axioms).

3. NORMATIVE PRINCIPLES OF SPECIAL METHODOLOGY

3.1. A sentence *n* is a normative principle of special methodology if it satisfies the following conditions:

3.1.1. *n* is a general (universal) norm (normative sentence) or rule applied in the research activity of at least one scientific discipline.

3.1.2. *n* is testable at least in the following way: if *n* is violated then either one of the first normative principles (cf. 1.4) is violated or one of the normative principles of general methodology (cf. 2.2.1–2.2.11 and 2.4.1–2.4.6) is violated.

3.2. Mathematics

3.2.1. Do not transfer laws from the finite to the infinite.

Since many of the laws are different for the finite and the infinite domain, a violation of the rule leads often to false results. Caution: Do not forget that there are interesting relations between the two domains which is substantiated by such theorems as the Compactness-Theorem and by results of Model-Theory.

3.2.2. Do not transfer laws from the infinitive to the finite. Justification: Same as in 3.2.1. Caution: Same as in 3.2.1.

3.2.3. Look for new mathematical structures which can be built (constructed) out of basic structures (such as groups) i.e. which are specifications of basic structures.

A violation of this rule leads to a violation of several normative principles of general methodology especially of principles 2.2.10, 2.2.11, 2.4.5 and 2.4.6. Caution: Do not stick to such structures. There are cases in which it may be more appropriate to investigate others.

3.2.4. If an equivalence relation is definable on some structure then try to construct all functions in such a way as to satisfy it.

If this rule is violated again several normative principles of general methodology are violated, especially principles 2.2.10, 2.2.11, 2.4.5, and 2.4.6. Caution: Same as 3.2.3.

Of the many further scientific disciplines two, physics and psychology, are taken as examples. One reason for that is that both are – compared with other disciplines – rather developed in respect to their methodological apparatus.

3.3. Physics

3.3.1. All physical laws should be invariant against coordinate-transformations.

A violation would violate the requirement for content, high degree of information or explanatory power, i.e. it would violate the first normative principles (cf. 1.4). Moreover principles 2.2.10 and 2.2.11 are violated too. Caution: Do not transfer such norms from physics to social or historical disciplines, where we do usually not have space-time-invariance.

3.3.2. All physical laws should be invariant against boundary (initial, random) conditions.[18]

If this rule is violated then so is the requirement for content, i.e. a first normative principle (1.4); moreover principles 2.2.10 and 2.2.11 are violated. Caution: Do not presuppose the concept of boundary condition (initial condition, random condition) to be sharply determined: Constants of nature (for example the proportion of the masses of proton and electron) have to be excluded.

3.3.3. All physical basic laws should be invariant against transformations in (relative to) all inertial reference frames. This is Einstein's law of Special Relativity.[19] It is a normative metanomological principle saying what every basic law of physics should satisfy.[20]

A violation would lead to a violation of the first normative principles concerning contentful and informative explanatory statements and of principles 2.2.10, 2.2.11, 2.4.5, and 2.4.6. Caution: Do not confuse the principle of relativity which is a normative law (rule) about laws with a law about physical systems. And observe further that it does not hold for any law but just for basic laws.

3.3.4. All physical basic laws should be invariant against transformations in any (moved) system. This is Einstein's law of General Relativity.[21]

Like in the case of the principle of special relativity a violation would violate the requirement of maximally contentful truth (approximate truth) and validity (approximate validity) – cf. 1.4 – and further it would violate the principles 2.2.10, 2.2.11, 2.4.5, and 2.4.6. Caution: Same as in 3.3.3.

3.3.5. When concerned with important theories (or hypotheses) then even the slightest deviations (which appear in the experiment) have to be taken into account and must be used to correct and improve them.

A violation of this rule – which is said to be due to Rutherford – would violate the important principles of general methodology 2.2.1 and 2.2.4. Caution: Slight deviations may sometimes result from a failure of the instruments.

3.3.6. "In experimental philosophy we are to look upon propositions collected by *general induction* from phenomena as accurately or very nearly true, notwithstanding any contrary hypothesis that may be imagined till such time as other phenomena occur by which they may either be made more accurate, or liable to exceptions."[22]

Although one cannot give a logical justification of a hypothesis arrived at by "general induction from phenomena" one can understand Newton's Rule IV as saying: Keep your original hypothesis as long as severe tests either change it into a more accurate version of falsify it. Stated thus the rule is subordinated to the important principle of general methodology 2.2.4. A violation of it would therefore violate also 2.2.4. Caution: Newton – at least in his famous work *Principia* – did not proceed by induction: He begins with definitions and axioms (principles of movement) and applies in fact the axiomatic method or the hypothetico-deductive method.

3.3.7. Continuity-Principle: Try to interpret the phenomena in such a way that they have always a continuous dependence upon their causes.

This rule is connected with a specific interpretation of rule 2.2.3 in the sense that 'mathematical structure' (of physical phenomena) is interpreted as *infinitesimal structure* or *structure to be described by differential equations*. A violation of the continuity-principle would therefore violate only a special form of the rule 2.2.3 which takes the infinitesimal structure as the only possible structure for physical phenomena (which in fact would be wrong).

Caution: The continuity principle is as a general normative principle invalid (unsatisfiable) since it is impossible to interpret all phenomena in such a way that they have always a continuous dependance upon their causes (Counter example: quantum phenomena).

3.3.8. The measurement of physical magnitudes (quantities) should be as undisturbed (trouble-free) as possible and if interference factors and observation errors occur they must be compensated and corrected to become as small as desired with the help of the error correcting calculus.

This principle is connected with the general principle 2.2.4. If it is violated, not the general principle 2.2.4 would be violated, but only a (unsatisfiable) specific form of 2.2.4 presupposing measurement situations in which every disturbance can be corrected (compensated) by the theory of error-calculation.

Caution: The rule 3.3.8 – though thought to be valid in parts of 19th century physics – is as a general normative principle invalid (unsatisfiable) since it is impossible to correct every disturbance to become as small as desired. Example for limitation: Heisenberg's uncertainty-principle.

3.4. Experimental Psychology[23]

3.4.1. Study all of the previous research-work that is relevant to the experiment.

This rule is a specification of the general methodological rule 2.4.2. It is also related to rules 2.2.5 and 2.2.1. Thus a violation of rule 3.4.1 would violate the rules of general methodology 2.4.2, 2.2.5 and to some extent 2.2.1. Caution: See the respective restriction to 2.4.2, 2.2.5 and 2.2.1.

3.4.2. State your hypothesis as explicitly as possible, preferably in mathematical or logical language.

This norm is an instance of 2.2.3 and is also related to rule 2.2.2. A violation would mean to violate 2.2.3 and partially 2.2.2. Caution: Don't construct empty symbolism without attempting to formulate the hypothesis first in natural language, using terms of the language of experimental psychology.

3.4.3. Define the variables stated in your hypothesis.

This rule is connected with rule 2.2.2 and is a special case of it. A violation would violate at least partially 2.2.2, even if there are other ways to make the concepts, used in a hypothesis (and in scientific approach in general), as precise as possible. Sometimes this rule is stated in an operationalistic form requiring that the respective variables are defined operationally.[24]

Caution: Do not presuppose anything which would even be logically impossible, for instance, that everything should be defined (without accepting primitives). Remember that measurements presuppose theories. Concerning operationalism one should not forget that laboratory operations refer only to observable objects and data, and that these are only a small part of reality, i.e. a small part of that which a hypothesis or theory speaks about.[25]

3.4.4. Specify the independent variable and the dependent variable on one hand and control the contaminating variables on the other.

Violating rule 3.4.4 means – under the assumption that variables have been already defined for the experiment and that it is an experimental test at all – that some necessary conditions are lacking for the fulfilment of the rule of general methodology 2.2.4. Caution: Don't search for absolute control and compensation (cf. rule 3.3.8) but try to make sure that no extraneous variable will affect different control-groups differently.

3.4.5. State the possible evidence reports for confirming or disconfirming your hypothesis.

If rule 3.4.5 is violated then trivially rule 2.2.4 and 2.4.4 are violated. Caution: Do not confuse 'confirmed' and 'true' and do not think that confirmation increases just with the quantity of positive instances (cf. rule 2.2.4).

4. NORMATIVE PRINCIPLES IN NORMATIVE SCIENCES

4.1. A scientific discipline is called *normative* iff it uses deductive nomological arguments for explanation, where the explanandum is a norm or a mixed proposition and where the premisses (the explanans) contain at least one law-like universal norm or mixed proposition.[26]

Examples of such normative disciplines are: ethics, pedagogics (education)[27] – if understood as explaining educational norms and confirming respective hypothesis, jurisprudence – if understood as being able to criticize rationally legal laws.

4.2. Examples of fundamental normative principles in normative sciences.

4.2.1. All fundamental (formal) ethical laws should be invariant against transformations in value-scales (value-systems). This normative principle was defended by Thomas Aquinas where he explains his formal basic principle for ethics: "The good should be done the bad avoided".[28]

This principle satisfies the rule 4.2.1 since it is independent of any value system. Other examples satisfying rule 4.2.1 are the most general laws of deontic logic.[29]

A violation of 4.2.1 would violate – analogously to a violation of the rule 3.2.1 in physics – the requirement for contentful and informative validity, i.e. it would violate the first normative principles (cf. 1.4). Moreover principles 2.2.10 and 2.2.11 would be affected too.

4.2.2. All legal laws should satisfy (a tolerant form of) the "ought-can" principle, i.e. the principle: "if p is obligatory (for a person to whom the law is addressed) then it must be possible to act in such a way that p is the case or that the state of affairs p is reached approximately."

If this rule is violated then normative principles which are unsatisfiable are allowed. And this again means that invalidity or meaninglessness can enter, since a norm which cannot possibly be satisfied, not even approximatively, is either invalid or meaningless. From this it is clear that a violation of 4.2.2 means a violation of the first normative principles (cf. 1.4).

4.2.3. All ethical laws should be consistent (compatible) with the most general laws of deontic logic.[30] Ethical laws not consistent with the most general laws of deontic logic are either inconsistent itself or can hardly be accepted as valid. Thus a violation of rule 4.2.3 would violate also the first normative principles (cf. 1.4).

University of Salzburg and
Institut für Wissenschaftstheorie,
Internationales Forschungszentrum Salzburg

NOTES

Concerning the examples of rules I am indebted to Professors Georg Kreisel (especially mathematics and physics) and Mario Bunge (general methodology and physics). The discussion of such rules in Bunge's books *Foundations of Physics* and *Scientific Research* was of particular value.

[1] K. R. Popper, 'The aim of Science', in K. R. Popper, *Objective Knowledge*, Oxford 1972, pp. 191–205, p. 24.
[2] C. S. Peirce, *Collected Papers*, ed. C. Hartshorne, P. Weiss and A. Burks, Cambridge, Mass. 1958–60; 2.135.
[3] Aristotle, *Metaphysics* I, 1: 980a21.
[4] I do not enter here into the difficult question how such general desires could be investigated or even tested. But there seems to be not sufficient reason to claim that scientific investigation is ruled out here.
[5] Cf. the discussion in Hintikka, *Knowledge and Belief*, New York 1962, p. 19 ff and condition C.K.
[6] The notion of approximate truth seems to originate in Popper's conception of versimilitude which he used in describing scientific progress. Cf. Popper, *Conjectures and Refutations*, London 1963, p. 228 ff. Cf. D. W. Miller, 'The Accuracy of Predictions', *Syntese* **30** (1975), pp. 159–191. The problem was later developed by R. Wojcicki; see his 'The Factual Content of Empirical Theories' in J. Hintikka (ed.), *Rudolf Carnap, Logical Empiricist*, Dordrecht 1975, pp. 95–122, and 'Semantic Conception of Truth in the Methodology of Empirical Sciences', *Studia Filozoficzne* 3 (1969), pp. 33–48. Cf. further: R. Hilpinen, 'Approximate Truth and Truthlikeness', in M. Przełęcki, K. Szaniawski and R. Wojcicki (eds.), *Formal Methods in the Methodology of Empirical Sciences*, Wrocław 1976, p. 19–42.
[7] For instance problems of existential import or problems like those, Wittgenstein raised in *Tractatus*, 5.53–5.5352.
[8] H. Putnam, 'There is at least one a priori truth', *Erkenntnis* **13** (1978), p. 153–170.

[9] A. A. Fraenkel, Bar Hillel and A. Levy, *Foundations of Set Theory*, Amsterdam, 1973, p. 294.

[10] D. Hilbert, 'Axiomatisches Denken', in D. Hilbert, *Gesammelte Abhandlungen*, 3 Vols. Berlin 1932-35, New York 1965; Vol. 3, p. 152.

[11] The versions (2), (3), (4) are – together with others – discussed more detailed in N. Rescher, *Many-valued Logic*, New York 1969, p. 144 ff.

[12] It has to be noted however that Putnam's claim is mainly the claim that "there is at least one a priori truth". Thus he wants to make his claim as weak as possible. On the other hand I see no reason to claim that the principle 1.2.624 is a priori in the sense that it is independent of experience since experience is involved in any understanding. Despite of that 1.2.624 can be generally acceptable in such a sense "that it would never be rational to deny it" (cf. Putnam, ibid., p. 155).

[13] For a detailed description of the Hilbert Program and its present status cf. G. Kreisel, 'Hilbert's Programme', in *Philosophy of Mathematics* (ed. P. Benacerraf and H. Putnam), Englewood Cliffs, N.J., 1964, p. 157-180.

[14] For a precise survey cf. A. Fraenkel, Y. Bar-Hillel and A. Levy, *Foundations of Set Theory*, Amsterdam 1973, ch. V, § 7.

[15] I. Newton, *Philosophiae Naturalis Principia Mathematica*, Book III, Rule I. In *I. Newtoni Opera quae extant omnia*, ed. S. Horsley, 5 Vols., London 1779-1785. Vol. II and III.

[16] Ibid., Rule II.

[17] Not only predictions in astronomy and natural science are meant here. A historical hypothesis may suggest ("predict") a place to find new historical sources or it may suggest to investigate the works of a historical person not yet considered for the explanation of a certain epoche. Cf. the conditions for a theory being nearer to the truth than another in Popper's *Conjectures and Refutations*, London 1963, ch. 10, p. 231 ff.

[18] Cf. K. R. Popper, 'A Revised Definition of Natural Necessity', in *Brit. J. Phil. Science* 18 (1967), p. 316-321. P. Weingartner, 'Sind die physikalischen Gesetze auf unser Universum beschränkt?', in R. Haller (ed.), *Philosophie und Physik*, Braunschweig 1975, p. 95-111.

[19] A. Einstein, 'Zur Elektrodynamik bewegter Körper', and 'Ist die Trägheit eines Körpers von seinem Energiegehalt abhängig?', in *Annalen der Physik* 17 (1905). Reprinted in H. A. Lorentz, A. Einstein and H. Minkowski, *Das Relativitätsprinzip*, Darmstadt 1958, pp. 26-50; 51-53; p. 26 and 51.

[20] Cf. M. Bunge, 'Laws of physical laws', in *American Journal of Physics* 29 (1961), p. 518. A basic law statement (like a field equation) is one containing no individual constants and entailing other general law statements (i.e. derived laws, like solutions of field equations). Cf. Bunge, *Foundations of Physics*, Berlin 1967, p. 213.

[21] A. Einstein, 'Die Grundlage der allgemeinen Relativitätstheorie', *Annalen der Physik* 49 (1916), reprinted in H. A. Lorentz, A. Einstein and H. Minkowski, *Das Relativitätsprinzip*, pp. 81-124; p. 83 and 86.

[22] I. Newton, *Philosophiae Naturalis Principia Mathematica*, Book III, Rule IV.

[23] For the rules stated here see F. J. McGuigan, *Experimental Psychology*, Prentice-Hall, 1968, p. 64 ff.

[24] Cf. McGuigan, *Experimental Psychology*, p. 27 and 65.

[25] For other presuppositions and confusions in operationalism cf. M. Bunge, *Scientific Research* I, New York 1967, p. 148 ff; *Foundations of Physics*, Berlin 1967, p. 27.

[26] A mixed proposition is a proposition which contains a statement which is true or false as one part and a norm which is valid or invalid as another part where both parts are connected with a logical connective. Examples of general forms: (1) "If p is the case then q should be (ought to be) the case" (cf. norms in criminal law). (2) "If p should be the case then q should be the case" (for instance: inferentially dependent norms). (3) "If p should be the case then q is the case" (for instance: "ought–can"-principle).

The problem that statements are true or false but norms may have other values is approached differently. Some transfer the usual truth-values to normative sentences (cf. Kutschera, *Einführung in die Logik der Normen, Werte und Entscheidungen*, Freiburg, 1973, p. 12 ff.), others say that the content of a normative sentence can be true or false. Another way – proposed by the author – is to introduce supervalues which are specified as true and false when applied to descriptive sentences, as valid and invalid when applied to normative sentences and left unspecified when applied to mixed sentences (cf. P. Weingartner, 'Über die Anwendung von Wahrheitswerten und Quantoren auf Normen', in *Strukturierungen und Entscheidungen im Rechtsdenken*, ed. I. Tammelo, Springer-Vienna 1978, p. 193–209). The problem need not concern us here. The important point is to realize that there are mixed propositions in the sense characterized above and that they play an important role in a normative explanation.

[27] Cf. P. Suppes, 'Can there be a normative Philosophy of Education?', in *Philosophy of Education 1968: Proceedings of the 24th Annual Meeting of the Philosophy of Education Society – Santa Monica*, Stanford 1968, p. 1–21.

[28] Cf. Thomas Aquinas, *Summa Theologica*, II, II, 94, 2. Observe the analogy between principle 4.2.1 and the fundamental principle 3.2.1 in physics.

[29] By "most general laws of deontic logic" I mean laws which are invariant in most of deontic systems like "if an action is forbidden the omission of it should be at least allowed". However very strong laws, holding just in one particular system are not candidates here.

[30] For the expression "most general laws of deontic logic", cf. note 29.

DAGFINN FØLLESDAL

EXPLANATION OF ACTION*

1. THE PROBLEM OF HOW TO EXPLAIN ACTIONS

Our actions are among the most important things in life; life consists in large part of actions. Action now has also become an important topic in philosophy. In the last 15 years alone about 600 papers on action theory have appeared in philosophical journals. A central question in many of these is: how can actions be explained? There is general agreement that actions are ordinarily explained by giving their reasons. But what is the connection between the reasons for an action and the action itself? When for example someone opens a window in order to let in fresh air, what is the connection between his reason for opening the window and his opening of the window? We do after all often say that we did this or that for this or that reason. Clearly we mean that there is a certain connection between reason and action such that giving the reason will contribute to an explanation of the action. But in what does this connection consist?

This question was taken up by Aristotle, who held that the connection is causal. A reason (e.g. for opening the window) can be viewed as consisting of two components, a desire (e.g. for fresh air), and a belief that the desire will be fulfilled by carrying out a certain action (in this example, opening the window). These two components together are the cause for the carrying out of the action, or, as Aristotle expresses this: "These two at all events appear to be sources of movement: appetite and mind." *De anima* III.10, 433ª 9–10. (Other passages in Aristotle show that he views the first component, the appetite or desire, as the actual cause, whereas mind or belief determines which of the desires is activated and to what type of action this desire then leads.)

2. DAVIDSON'S VIEW ON EXPLANATION OF ACTION

A series of objections have been raised against Aristotle's view. The most important were put forth at the end of the fifties and the beginning of the sixties by Miss Anscombe, H. L. A. Hart, Anthony Kenny, and several other British and American philosophers. Aristotle's view was for all

231

R. Hilpinen (Ed.), Rationality in Science. 231–247.
Copyright © *1980 by D. Reidel Publishing Company.*

intents and purposes given up when in 1963 Donald Davidson, in the article 'Actions, Reasons and Causes' (Davidson, 1963), gave a defense of it. He provided a new precise formulation of and argument for the view. Davidson's article became one of the most widely read of the 600 papers that I mentioned at the outset. It has inspired many of the other articles; in addition it is frequently referred to and has been re-printed in a number of anthologies. Davidson's argument has not been refuted and many hold that Davidson has overcome the objections that have been raised against Aristotle's conception. In the following I will discuss Davidson's argument. I shall declare myself in agreement with pretty well all of Davidson's points; in particular, his observation that the reasons for action may well be causes of action. However, I will try to demonstrate that the argument has one fatal weakness that under-mines his view on explanation of action.

Let me, however, first outline Davidson's argument. One of the main objections against viewing reasons as causes is that there appears to be no law-like connection between reason and action. On one occasion, for example, my desire for fresh air, together with my belief that I will get fresh air by opening the window, is a reason for my opening of the window. But on another occasion I have the same desire and the same belief, but I do not open the window, perhaps because there is someone in the room who is sensitive to draft. And how can it make sense to say that *A* is a cause of *B*, when now and again *A* occurs without *B*? The problem is Hume's old problem, which led him to define cause in the following manner:

We may define a cause to be 'An object precedent and contiguous to another, and where all the objects resembling the former are placed in like relations of precedency and contiguity to those objects, that resemble the latter'.[1]

Davidson accepts this definition of cause and admits that while laws are certainly necessary for causal explanations, it does not seem that there is a law-like connection between reason and action. How then can he claim that reasons are causes?

Davidson's argument rests on the observation that there is an intimate connection between description and causality that the philosophers who have worked on these problems have evidently failed to notice:

Every entity or event, every reason and every action, can be described in a number of different ways. That I open the window, can for example also be described in a purely physiological manner, namely as a contraction of certain muscles etc. Causal laws have the form "every event of type t_1 has an event of type t_2 as successor". Depending now on the manner in which I describe the two events, they will instantiate or fail to instantiate a given causal law. If the one event is described as t_1 and the other as t_2, then they instantiate the above law. But if the events are described in some other way, if, that is, we focus on other properties or aspects of the events, then they do not instantiate the law.

This means that we cannot say that two events in themselves instantiate a causal law, but rather that it is events described in a certain manner that instantiate laws. To return to our example of the opening of the window: when I describe my reason and my action in the usual manner, they do, as we saw, not instantiate a causal law. It seems possible to find counterexamples to every general law that purports to connect reason and action.

The fact that my reason and my action do not instantiate a law when they are described in the usual manner does not prevent them from instantiating a law when they are described in another manner, e.g. physiologically. A crucial part of Davidson's argument is now that he holds an "identity theory" of the mental and the physical. He maintains that mental events, e.g. that one believes or wants something, are also physiological events. That is, a mental event is always merely an aspect of an event that also has a physiological aspect. (This does not mean that similar mental events are always similar in physiological respect, and conversely, says Davidson. It is quite possible that two events that have completely similar mental aspects, e.g. correspond to exactly the same desire in us, have different physiological aspects. Likewise, two events which have completely similar physiological aspects may have different mental aspects, according to Davidson. There are no regularities connecting the mental and the physical.)

Davidson does not just assume his identity theory, he argues for it. His argument makes use of the same basic ideas that we shall encounter in his argument that reasons are causes, and although I am not going to take up his identity theory for criticism in this paper, it may be useful to review it very briefly.

(i) Mental events may bring about physical events, and conversely. E.g. light rays within my visual field may bring about beliefs in me, and beliefs and desires in me may bring about actions, which have physical aspects.

(ii)   In order for one event to bring about another they have to be connected by a strict law.

(iii)   There are no strict laws connecting events described in a mental vocabulary with one another nor connecting them with events described in a physical vocabulary.

Given (i) and (ii), it follows that mental events and physical events must be describable in such a way that under these descriptions they instantiate strict laws. In view of (iii), these strict laws have to be formulated exclusively in a physical vocabulary. Consequently, in order to be able to bring about physical events, mental events must be describable in a physical vocabulary, i.e. have physical aspects. Hence the identity theory is proved.

This argument of Davidson's is valid in the sense that the conclusion follows from the premisses. However, I do not regard it as a sound argument, since, depending on what one meant by a strict law, I regard either premiss (ii) or premiss (iii) as false. I shall give my arguments for this in the last section of this paper.

Let us now return to Davidson's view on explanation of action. Davidson's next move is to claim that when we say A is a cause for B, we do not require there to be a causal law which is instantiated by A and B under the descriptions 'A' and 'B'. All that is needed, is that there be some causal law and some descriptions of A and B such that A and B under these descriptions instantiate the causal law.

On this point there is an ambiguity in Hume, Davidson notes. Hume's definition of cause, which was cited above, could also be formulated as follows: A is a cause of B if and only if A under the description 'A' and B under the description 'B' instantiate a causal law. But Hume's own formulation, as well as his arguments, suggest rather that the first version of the definition of 'A is a cause of B' is that intended by Hume. That is, the causal law that connects A and B need not be exemplified by A and B under the descriptions 'A' and 'B'.

We thus see that there is a characteristic difference between a case in which A and B instantiate a causal law, and a case in which A is a cause of B. A and B instantiate a causal law *only under certain descriptions*, whereas A is a cause of B, *independently of how A and B are described*.

In Frege's terminology, Davidson's argument as presented above may be expressed as follows: In assertions that A and B instantiate a causal law, the names 'A' and 'B' are used indirectly; that is, the manner in which we refer to A and B determines whether the assertion is true or false. In assertions of causality of the form 'A is a cause of B', on the other hand, the names 'A' and 'B' are used directly: how A and B are referred to is irrelevant. Or, in Whitehead and Russell's terminology: causal statements are transparent, while assertions that A and B instantiate causal laws are opaque. This contrast between causal laws and assertions of causality, together with the identity theory of the mental and the physical, constitute the gist of Davidson's argument.

If, with Davidson, we hold that reasons and actions, under physiological descriptions, instantiate causal laws, then there is hence no obstacle in the way of viewing reasons as causes of actions. To be sure, we do not know the causal laws that connect reasons and actions. However, as Davidson observes, this is also the case for standard causal explanations. We are often convinced that A is a cause of B, but we are not in a position to formulate any causal law that connects A and B; we are nevertheless convinced that such a law exists. Davidson says that it is a mistake to believe that we have given no explanation unless we have provided a causal law. Certainly a suitable law will be required in order to make the explanation complete, but even without such a law, the assertion that A is a cause of B can contribute to an explanation of B. Herewith Davidson concludes his argument.

3. CRITICISM OF DAVIDSON'S VIEW

I will now turn to my criticism of Davidson's view. I have presented in summary form his whole argument and I have expressed my agreement with his observation that causal statements are transparent while assertions that causal laws are instantiated are opaque.

I therefore agree completely with Davidson that even if there were no laws that are instantiated by reasons when described as reasons and actions when described as actions, reasons might still be causes of actions, since reasons and actions might instantiate causal laws under other descriptions.

However, this would only prove the following:

(1)   Reasons *may be* physical causes of actions

One would need something more in order to establish:

(2)   Reasons *are* physical causes of actions

For example, one might add premisses (i), (ii) and (iii) above, which Davidson accepts.

Even more, however, is needed in order to explain actions by appeal to reasons. In the following, I shall argue for two theses:

(3)   The assumption that reasons are physical causes of actions is not *sufficient* for reason explanation to work.

(4)   The assumption that reasons are physical causes of actions is not *necessary* for reason explanation to work.

While there is no doubt that Davidson accepts (3), it seems fairly clear that he does not accept (4). My main disagreement with Davidson therefore concerns (4). However, this disagreement is a reflection of a disagreement relating to (3). Davidson, it seems, thinks that his view that reasons are physical causes of actions contributes more to reason explanation than I think it does. It will be helpful to consider this disagreement first before going on to (4). Let us therefore consider (3).

The crucial reason why (3) is true is that in explanations names are used indirectly. If I explain B by citing A, if I say "B occurred for the reason A" or "B occurred because of A", then the manner in which I refer to A and B determines whether what I say can function as an explanation.

To spell this out more in detail: If one asks, "Why did B occur?", not every reply of the type "B occurred for the reason A", "B was caused by A", or "A was a cause of B" will be an explanation, even if the answers are factually correct; that is, even if there is in fact a causal law and descriptions of A and B such that A and B under these descriptions instantiate the causal law.

To say that A is a cause of B does not contribute to an *explanation* of the occurrence of B unless there is a law which is instantiated by A and B under *approximately* these descriptions. (By 'approximately' I mean that the descriptions 'A'' and 'B'' under which A and B instantiate the law must be such that the person for whom the explanation is intended believes that 'A'' denotes everything that 'A' denotes, and that 'B' denotes everything 'B'' denotes.) If the only law that connects A and B is instantiated by A and B only under descriptions that differ radically from 'A' and 'B', then the assertion that A is a cause of B will be more the pronouncement of an oracle than it will be an explanation. The fact that '*A*' and '*B*', in assertions such as "*A* was a cause for *B*", is used directly, should not mislead us into forgetting or overlooking the fact that in explanations names are used indirectly. The descriptions '*A*' and '*B*' that are used in explanations must not be too distantly or coincidentally related to the descriptions under which *A* and *B* instantiate the law. (This constraint on descriptions applies in my view to all types of explanation, be it explanation by means of reasons or by means of causes.)

Now, according to Davidson the descriptions that are used in explanations by means of reasons and those that are used in causal explanations differ considerably:

> The laws whose existence is required if reasons are causes of actions do not, we may be sure, deal in the concepts in which rationalizations must deal. If the causes of a class of events (actions) fall in a certain class (reasons) and there is a law to back each singular causal statement, it does not follow that there is any law connecting events classified as reasons with events classified as actions – the classifications may even be neurological, chemical, or physical. (Davidson 1963, p. 699)

Therefore, if I am right in what I have just said about the constraints placed on descriptions in explanations, Davidson's claim that reasons are physical causes does not answer the question: Why can explanations based on reasons function as explanations? This means that to the extent that our understanding is broadened by coming to know the reason for an action, this broadening must be attributed to a connection between reason and action other than a causal connection of the type discussed by Davidson.

Davidson is quite aware that names are used indirectly in explanation contexts. It is obvious that one cannot explain an action by just mentioning some desire or belief that the agent has and expressing one's conviction that some causal law or other connects this desire or belief with the action. There must be some regularity, which we can at least roughly specify, that connects the reason with the action. Davidson thinks that there are such rough regularities, but maintains that they cannot be refined into the kind of strict laws that are needed for causal connections (cp. premise (iii) above).

Davidson gives two reasons for this. One is that

> What emerges, in the *ex post facto* atmosphere of explanation and justification, as *the* reason frequently was, to the agent at the time of action, one consideration among many, *a* reason. Any serious theory for predicting action on the basis of reasons must find a way of evaluating the relative force of various desires and beliefs in the matrix of

decision; it cannot take as its starting point the refinement of what is to be expected from a single desire. The practical syllogism exhausts its role in displaying an action as falling under one reason; so it cannot be subtilized into a reconstruction of practical reasoning, which involves the weighing of competing reasons. (Davidson 1963, p. 697)

Davidson's second reason for thinking that there cannot be any strict laws that are instantiated by reasons described as reasons and actions described as actions, is that this follows from Quine's thesis of indeterminacy of translation. However, although Davidson has contended this in several of his writings, he has never presented his argument that this is so. It is not obvious how this argument is supposed to run. As long as the argument has not been spelled out, it is hard to evaluate it and we have to depend on the first of the reasons that Davidson gives, in the first passage quoted above. Let us therefore turn to it.

Davidson's first reason for holding that there are no strict laws instantiated by reasons described as reasons and actions as actions can be illustrated by the example concerning opening the window which was mentioned at the beginning of this paper. This example was there used to illustrate one of the main objections against viewing reasons as causes. Davidson surmounted it by pointing out that the reason and action may instantiate strict laws when they are redescribed in physical terms. However, he still seems to think that it shows that there can be no constant correlation between reasons when described as reasons and actions when described as actions. So let us consider it.

A person may have a pro attitude towards actions with a certain property (in our example, opening the window, which will provide fresh air), it may be within his power to perform an action with that property, he may believe that this is within his powers, and at one time he may perform the action and even give the pro attitude and the belief that the action had this property as his reason for performing it. And yet, at some other time he may have the same pro attitude and the same true beliefs and fail to perform the action. Admitting all this, can we avoid the conclusions that there is no regularity connecting reason and action? Or how can we in the face of all this still claim that there is such a regularity?

The answer, I think, that the regularity must take into account all the different ways in which the action and its consequences can be described. One and the same action may under one description have a property that the agent dislikes. In giving our reason for performing an action, we may state one or more of our dominant pro attitudes and beliefs concerning the action, but our decision whether or not to perform the action is usually based on considerations and weighing of many more features of the action and its consequences, of the likelihood of the action actually having these features and consequences, and on comparison of this action with other alternative actions that we believe ourselves free to perform in the given situation. Often strong pro attitudes and beliefs will be outweighed by other even more weighty contra attitudes and beliefs concerning other features and consequences of the contemplated action or by strong pro attitudes and beliefs concerning the features and consequences of some alternative action.

The regularities that come into play are therefore much more complicated than those that involve just one reason. They are, as also Davidson hints at in the last passage quoted above, those that one seeks to clarify in decision theory. One there seeks to find a way of evaluating the relative force of the agent's various desires and beliefs in what Davidson calls "the matrix of decision." Such a decision theoretic approach to explanation of action has been proposed by several people, including Dray (1957), Hempel (1961–62) and Stegmüller (1969). In a way, Davidson favors it, too, but he thinks that it cannot do the job alone; it has to be supplemented with a view of reasons as physical causes of actions.

Admittedly, there are many qualifications that have to be made here. As observed by Hempel (1961–62) and Stegmüller there is first the problem that decision theory does not always give us a unique answer. There may be ties between equally good alternatives. What is worse, in the case of decisions under uncertainty, we still have no generally accepted view as to what is the rational thing to do. Also in most practical applications of decision theory we do not have sufficient information about what courses of action are being considered, what consequences are taken into account and what the probabilities and values of these consequences are supposed to be.

Secondly, decision theory is normative, it does not tell us what people that have given desires and beliefs do, but what they ought to do. As Hempel has pointed out (1961–62), one therefore cannot, as Dray (1957), just use the normative theory to explain why people act the way they do. However, Hempel's alternative suggestion also has its problems. Hempel proposes that one should supplement Dray's account with two empirical hypotheses, one to the effect that the agent is rational and one to the effect that every rational agent does what he ought to do, i.e. he does

what is rational according to decision theory. Are these premises empirical hypotheses, or what are they? Can one first attribute certain
desires and beliefs to an agent and then ask whether he is rational? The
difficulty is one that has been brought out very forcefully by Donald
Davidson in several other of his writings: attributions of rationality and
attributions of desires and beliefs have to go hand in hand.

The relation between rationality, desire and belief is in fact quite
complicated. A person's actions are not the only clues we have to his
beliefs and desires, even if we include among his actions also the actions
in which he uses language and says or writes something. For example,
observations of what reaches his sensory surfaces are pertinent, in the
light of our theories of how beliefs and desires are formed and changed.
Here, too, a notion of rationality comes in: what beliefs and desires
are it rational to have? To what extent are our beliefs and desires formed
by processes that we consider rational and to what extent are they formed
by other, non-rational processes?

Similarly in action theory, only few of our actions are the result of
rational deliberation, often one acts on impulse, in panic, etc. Also, there
may be factors that are quite important for the outcome, factors that
one knows, but temporarily forgets or overlooks. In order to explain
actions, it is important to know just what beliefs and desires were entertained by the agent at the time of action. In particular, one must know
what alternatives were considered and what consequences were taken
into account. There is often very little evidence to go on here, and
one has correspondingly much leeway for alternative explanations. In
order to prevent one's explanation from becoming vacuous, one must
therefore support by evidence and argument these hypotheses concerning
what the agent believed and desired and what alternatives and consequences he considered. Thus, for example, if one purports to explain why
an agent did what he did by claiming that he was unaware of an important alternative, this claim must be made plausible, perhaps by
pointing out that some crucial information never reached him.

Another issue has to do with how much we should include among
the beliefs and desires that a person has and that determine his actions.
While in explicit deliberation one normally takes into account only
beliefs and desires that are consciously entertained at the time, most
practical decisions clearly also depend on beliefs that are not consciously
entertained but can be elicited by various means (cf. e.g. Hempel,
1961–62). Davidson, Suppes and Siegel (1957) have studies choices which

seem to depend quite systematically on subjective preferences and utilities that the subject is not aware of. Also, there are actions that apparently can best be explained by attributing to the agent beliefs and desires that it is very hard to elicit by any method and that the agent might deny that he has. A simple example presented by Suppes in an unpublished paper is the following: A young boy who has just entered puberty and has an attractive female teacher very frequently comes up to the teacher after class to ask questions concerning the school work. When asked why he does it, he may answer that he has these questions concerning the school work and that he wants to learn. This may be his sincere answer, but we may want to give a different explanation of his actions. Similar examples abound in psychoanalytic case studies, in experiments with hypnosis, etc.

Note, by the way, that this last example shows that normative decision theory is not normative in the sense that rational explanation always takes precedence over other, more causal kinds of explanation. Rather, the situation is the following: We always start our explanation of action from within our current psychological, physiological and other theories of man. When on the basis of these theories we have no reason to think that a person's behavior springs from causes that are not reasons or from unconscious motives, we shall assume that he acts as a rational agent in the sense of decision theory.

One further remark on normative decision theory: Just as in our theory of how we form beliefs we have to temper our pure rational account with non-rational deviations, so in our theory of decision making non-rational deviations have to come in. Tversky and Kahneman (1980), for example, have pointed out that seemingly inconsequential changes in the formulation of choice problems cause significant shifts of preference. These violations of rationality resemble the effects of changing perspective on perceptual appearance and consequently on belief. Kahneman and Tversky also observe that such "framing" effects, like perceptual illusions, often remain appealing even when they are recognized as mistakes. Cp. also the various other empirical studies mentioned in the bibliography, by Tversky, Davidson and their collaborators. See also the discussion between Cohen, Kahneman and Tversky. Generally, it seems to me that such empirical studies of practical decision making are just as relevant to the explanation of action as empirical studies of perception and reasoning are relevant to our attributions of beliefs and desires to persons.

We are still far from a satisfactory theory of how to explain actions by appeal to reasons. Many problems remain. However, these problems have to be faced by Davidson, too. In spite of all that he says about reasons being physical causes, the only arguments he can give for holding that one reason rather than another is a reason for an action depend on decision theoretic considerations. The appeal to physical causes contributes not the slightest to help us decide what the reasons for an action are. Davidson is no doubt aware of this. The only reason he gives for bringing in physical causes, is that he thinks they are needed to account for the full force of 'because' in "the agent did B because of A". However, do we need laws of the kind commonly referred to in physics in order to account for the 'because'? To answer this question let us see how reason explanation compares with the kind of causal explanation we are accustomed to in physics.

4. REASON EXPLANATION COMPARED WITH EXPLANATION IN PHYSICS

We have noted that in reason explanation there is never just *one* reason, i.e. one desire or belief, that enters into the explanation. However, as in other areas of explanation, when an explanation is given, we usually pick out one or just a few factors, that we give as *the* reason(s) for action. Which factors we pick depends upon their relative weight and the circumstances of the explanation, for example, what factors are already known to the person for whom the explanation is intended.

Hence, there seems to me to be no incompatibility between there being a number of reasons that enter into explanation by reason and our still giving only one, or a small number of these as *the* reason(s) for the action. In fact, there seems to be a complete parallel here with the situation in causal explanation. In the case of causal explanation, too, one often just mentions one, or a small number of the causes of an event, e.g. those of which the person for whom the explanation is intended is not yet informed.

I think that the parallelism between explanation by reason and explanation by causes extends far beyond this. I will even suggest that rather than assimilating reasons to causes in the simple way that the instantiation of causal laws suggests, one should be aware that causal explanation, like explanation by reason, makes use of a whole intricate theory and not a single simple causal law. To take an example, let us consider the case where we explain why an iron ball is falling towards the center of the earth by saying that the ball is heavy and that we have a law of physics to the effect that every heavy body here on earth will fall towards the center of the earth. Clearly, just as in our opening the window example, this is sometimes true, sometimes false. If some other factor enters the picture, like a support that keeps the object in place, an electromagnetic field or the like, the ball will perhaps not fall, perhaps it even will move upwards.

We could of course save our law in the face of counter-examples like this by adding clauses to it, we might say: a heavy object here on earth which is not supported and not in a magnetic field will fall towards the center of the earth. Similarly in the case of actions, we might in our example concerning the opening of the window begin with the following simple 'law': A person who wants fresh air and believes that he will get it by opening the window, will open the window. Noting the many exceptions from this 'law, we might improve it to: "A person who wants fresh air and believes that he will get it by opening the window and who also does not believe that anybody in the room will suffer from the draft, will open the window". However, both in the example from physics and the example from action theory, our new revised 'law' has exceptions. They both have to be supplemented with ever new clauses.

In action theory, we quickly see that there is little point in trying to formulate ever more complicated laws. Instead, we attribute to each agent a number of propensities towards action, a set of desires and beliefs, and we devise a decision theory which tells us how in a given situation all these propensities are fused into one resultant propensity for action. Similarly in physics we attribute to each object a number of propensities, mass, electric charge, a certain position relative to other masses and electric charges, etc., and on the basis of this we determine the resultant propensity for e.g. movement in the given situation. Another way of looking at this is that physical objects and events, like actions, can be described in numerous ways, for each of which they instantiate different 'laws', i.e. show different propensities. These laws, or propensities, by the way, need not be deterministic. They may be probabilistic in action theory as well as in physics.

Formally, there are hence many similarities between the situation in physics and that in action theory. The way I see it, reasons that fit into the explanatory pattern of rational explanation thereby qualify as causes, just as physical events that fit into the pattern of physical explanation thereby qualify as causes. They both account equally well for the 'because' in contexts of the type "*B* happened because of *A* ".

5. CONSEQUENCES OF THIS DISCUSSION FOR DAVIDSON'S ARGUMENTS

Our discussion of thesis (3) above has not only served as an argument for that thesis. It also has delivered as a by-product an argument for thesis (4). I have argued that reasons need not be redescribed in physical terms in order to account for the 'because' in "*B* happened because of *A* ". If I am right, it is not necessary to assume that reasons are physical causes of actions in order to make reason explanations work. So, contrary to Davidson, I maintain that thesis (4) is true.

Not only Davidson's view on the explanation of action, but also his argument for the identity theory is undercut by these observations. Either premiss (ii) or premiss (iii) or both have to be given up. If by 'strict law' Davidson means the traditional kind of causal law of the form "every event of type t_1 has an event of type t_2 as successor", then, as I have just argued, premiss (ii) is false. Such law-like connections are not required in order that one event shall bring about another. As we have seen, even in physics such laws should give way to talk about dispositions.[2] If on the other hand, Davidson by 'strict law' means more complex regularity patterns of the kind found in physics and also in decision theory, then premiss (iii) is false, or in any case it is un-supported by the arguments Davidson has given so far. It is in fact significant that the argument from indeterminacy of translation that Davidson so often refers to, is left so vague that it is not even clear whether it applies to regularities of this latter, decision theoretic kind, or only to simple one-factor laws of the traditional kind.

6. CONCLUSION

My assimilation of reason explanation to causal explanation, or rather of causal explanation to reason explanation, is not intended to belittle the differences between the study of man and the study of the physical world. Let me end my paper by pointing out three such differences:

First, in physics one has found a way of reducing the basic pro-pensities of physical objects to very few, mass and electric charge, position in gravitational and electromagnetic fields etc., and one hopes for further reductions. In action theory, the situation is much more complicated. We have to deal with values, like money, beauty, unspoiled nature, peace, love, etc., that we have not learned how to compare and that are perhaps incomparable. Also, empirical studies indicate that the weight given to these factors and the manner of their interplay may vary from person to person.

Another factor which also leads to formal differences between physics and action theory, is the capacity that the objects we study in action theory, viz. human beings, have to consider various possibilities, some of which will better satisfy our preferences than those we would reach if we only followed a gradient from where we are to some local maximum.

Finally a third, and most important factor which has to be taken into account in action theory and which, as far as we know, has no counterpart in physics, is our capacity, as human beings, to perform actions, or acts, of reflection upon our own situation and what happens to us, and to let this reflection work back on our beliefs and our values and thereby in turn on our actions. This is what in cybernetics one has sought to capture by mechanisms of feedback etc., but so far, it seems that we here have a feature of man that is not easily imitated by physical processes. This is not to belittle cybernetics. One of the main difficulties in discussing self-reflection and self-consciousness is to characterize these notions more exactly, to understand better what they consist in. Here, cybernetics, by pressing for sharp questions and sharp answers, has helped us to see better what these capacities, which seem to be so characteristic of man, are and what they are not.

University of Oslo and
Stanford University

NOTES

* I am grateful to Miss Elizabeth Anscombe for her comments on an earlier version of this paper, that was read at a meeting on 'Explanation', organized by Stephan Körner in Bristol 1973, where Miss Anscombe served as a commentator. An earlier version of this article was printed in German, under the title 'Handlungen, ihre Gründe und Ursachen', in Hans Lenk, ed., *Handlungstheorien - interdisziplinär*, Fink Verlag, Munich, 1979, Vol. 2, pp. 431–444. I thank Professor Lenk and Fink Verlag for their permission to publish this English version of the paper in this volume.

I also wish to thank Lars Bergström, Michael Bratman, Risto Hilpinen and John Perry for helpful criticism.

[1] David Hume (1739), page 170 of the Selby-Bigge edition. Quoted by Davidson (1963), pp. 696-697.
[2] For a similar observation, see Nancy Cartwright 1980a and 1980b.

REFERENCES

Brandt, R., and Kim, J.: 1963, 'Wants as explanations of actions', *Journal of Philosophy* **60**, pp. 425-35.

Cartwright, N.: 1980a, 'The truth doesn't explain much', *American Philosophical Quarterly* **17**, pp. 453-457.

Cartwright, N.: 1980b, 'Do the laws of physics state the facts?' *Pacific Philosophical Quarterly* **1**, pp. 75-84.

Cohen, L. J.: 1979, 'On the psychology of prediction: whose is the fallacy?', *Cognition* **7**, pp. 385-407. Reply by Kahneman and Tversky, pp. 409-11, rejoinder by Cohen, *Cognition* **8** (1980), pp. 89-92.

Davidson, D.: 1963, 'Actions, reasons, and causes', *Journal of Philosophy* **60**, pp. 685-700 (reprinted in Brand, M., ed., *The Nature of Human Action*, Scott, Foresman and Company, Glenview, Ill., 1970, pp. 67-79 and in several other anthologies).

Davidson, D.: 1970, 'Mental events', in Foster, L., and Swanson, J. W., eds., *Experience and Theory*, The University of Massachusetts Press, Amherst, 1970, pp. 79-101.

Davidson, D.: 1974, 'Psychology as philosophy', in Brown, S. C., ed., *Philosophy of Psychology*, Macmillan, London, pp. 41-52.

Davidson, D.: 1975, 'Thought and talk', in Guttenplan, S., ed., *Mind and Language* (Wolfson College Lectures 1974), Clarendon Press, Oxford, pp. 7-23.

Davidson, D.: 1976, 'Hempel on explaining action', *Erkenntnis* **10**, pp. 239-53.

Davidson, D.: 1978, 'Intending', in Yovel, Y., ed., *Philosophy of History and Action: Papers Presented at the First Jerusalem Philosophical Encounter*, Reidel, Dordrecht, and the Magnes Press, The Hebrew University, Jerusalem.

Davidson, D., and Marschak, J.: 1959, 'Experimental tests of a stochastic decision theory', in Churchman, C. W., and Ratoosh, P., eds., *Measurement: Definitions and Theories*, Wiley, New York, pp. 233-69.

Davidson, D., Suppes, P., and Siegel, S.: 1957, *Decision Making: An Experimental Approach*, Stanford University Press, Stanford.

Dray, W.: 1957, *Laws and Explanation in History*, Clarendon Press, Oxford.

Dray, W.: 1963, 'The historical explanation of actions reconsidered', in Hook, pp. 105-35, and in Gardiner, pp. 66-89.

Gardiner, P.: 1974 (ed.), *The Philosophy of History*, Oxford University Press, Oxford, 1974.

Hempel, C. G.: 1961-62, 'Rational action', *Proceedings and Addresses of the American Philosophical Association* **35**, pp. 5-23 (reprinted in Care, N. S., and Landesman, C., eds., *Readings in the Theory of Action*, Indiana University Press, Bloomington, 1968, pp. 281-305).

Hempel, C. G.: 1963, 'Reasons and covering laws in historical explanation', in Hook, pp. 143-63, and in Gardiner, pp. 90-105.

Hempel, C. G.: 1965, *Aspects of Scientific Explanation*, Free Press, New York.

Hook, S.: 1963 (ed.), *Philosophy and History: A Symposium*, New York University Press, New York.

Hume, D.: 1739, *A Treatise of Human Nature*, Selby-Bigge edition, Clarendon Press, Oxford, 1978.

Stegmüller, W.: 1969, *Probleme und Resultate der Wissenschaftstheorie und Analytischen Philosophie, Band I, Wissenschaftliche Erklärung und Begründung*, Springer, Heidelberg/ New York.

Tversky, A.: 1972a, 'Choice by elimination', *Journal of Mathematical Psychology* **9**, pp. 341-67.

Tversky, A.: 1972b, 'Elimination by aspects: a theory of choice' *Psychological Review* **79**, pp. 281-300.

Tversky, A.: 1975, 'A critique of expected utility theory: descriptive and considerations', *Erkenntnis* **9**, pp. 163-73.

Tversky, A. and Kahneman, D.: 1974, 'Judgment under uncertainty: Heuristics and biases', *Science* **185**, pp. 1124-31.

Tversky, A.: 1979, 'Prospect theory: an analysis of decision under risk', *Econometrica* **47**, pp. 263-91.

Tversky, A.: 1980, 'The Framing of decisions and the rationality of choice', unpublished manuscript, Stanford.

Tversky, A. and Sattath, S.: 1979, 'Preference trees', *Psychological Review* **86**, pp. 542-73.

INDEX OF NAMES

Anaxagoras 207
Anscombe, E. 231, 245
Aquinas, T. 227
Åqvist, L. 16
Aristotle 33, 82, 156, 161, 204, 210, 231, 232

Bach, J. S. 164
Balzac, H. de 163
Bateman 174
Bayes, T. (Bayesian) 171, 177–180, 184, 188
Becquerel, H. 174
Beethoven, L. van 164
Bell, J. S. 176, 177
Bernoulli, D. 177
Bochvar 215
Bohr, N. 153, 155
Bolzmann, L. 156, 157, 159
Boyd, R. 92
Brecht, B. 86, 133
Broda, E. 157
Bruno, G. 160
Byron, Lord G. 161

Carnap, R. 25, 67, 137, 139
Cézanne, P. 164
Chisholm, R. M. 15, 22–24
Clauser, J. F. 177
Cohen, L. J. 241
Copernicus, N. 159
Craig 96
Curie, P. and M. 174

Dante, A. 164
Darwin, C. 159, 160, 164
Davidsson, D. 232–242, 244
Deese, J. 35
Dempster, A. P. 184
Descartes, R. 13, 203, 206, 207

Diodorus 205
Dostoievski, F. M. 164
Dray, W. 239
Ducasse, C. J. 25
Dummett, M. 97, 108

Eccles, Sir J. C. 159
Einstein, A. 105, 164, 175, 176, 212, 219, 224
Erlanger, J. 67, 74
Euclid (Euclidean) 103, 152, 204, 206
Eudoxus (of Knidos) 204

Feyerabend, P. 1, 2, 67, 91
Fillenbaum, S. 36
Finetti, B. de 177, 179, 185
Fourier, J. B. 207
Freedman, S. J. 177
Frege, G. 234

Galileo Galilei 151, 154, 160
Gauthier, D. 4
Geach, P. T. 13
Geiger, H. 86, 174
Gentzen, H. 216
Gibson, Q. 7, 8
Good, I. J. 184
Goodman, N. 36
Gödel, K. 216
Grundtvig, S. 131

Hart, H. L. A. 231
Harré, R. 92
Heisenberg, W. 153, 155, 159, 160, 225
Hempel, C. G. 4–7, 29–31, 33, 239, 240
Hesse, M. 92
Hilbert, D. 214, 216
Holt, R. A. 177
Horne, M. A. 177
Hume, D. 13, 155, 232, 234

249

Ibsen, H. 162
Imlay, R. A. 23, 24

Jacobi, F. H. 207
James, W. 18
Jeffrey, R. 31, 32, 33

Kahneman, D. 241
Kanger, S. 16
Kant, I. 66, 75, 81, 82, 155, 173, 177
Katz, J. 35
Kenny, A. 231
Kepler, J. 154, 206
Kleen, D. 215
Klein, F. 207
Kohlrausch, K. W. F. 174
Kolakowski 84
Koopman 179
Kuhn, R. 67, 91

Lakatos, I. 67
Laplace, H. 172–174, 176, 177
Lehrer, A. 36
Lehrer, K. 196
Leibnitz, G. W. von 65
Lessing, G. E. 129
Levi, I. 14, 18, 24, 25
Lorentz, K. 105
Luce, R. D. 3
Lukasiewicz, 215
Lyons, J. 38

Maxwell, J. C. 155
Mendel, J. G. 155, 196
Merton, R. K. 197
Michelangelo 164
Miller, G. 36, 37
Monod, J. 159
Mozart, W. A. 164

Naess, A. 37
Neumann, J. von 176
Newton, I. 101–104, 152–156, 164, 173,
    213, 219, 225
Nietzsche, F. W. 87
(Notwen-Theory) 102–104

Ockham, William of 192, 219
Osgood, C. 35

Parmenides 203
Peano, G. 216
Pearson 41, 46
Peirce, C. S. 66, 71
Picasso, P. 164
Planck, M. 155
Plato 203–205
Podolsky, B. 176
Poisson, S. D. 174
Popper, K. R. 2, 67, 92, 98, 155, 209
Pörn, I. 16
Post, J. 215
Ptolemäus 206
Putnam, H. 92, 95, 213, 215
Pythagoras 204

Quine, W. V. O. 36, 42, 46–48, 91,
    93–95, 97, 159, 238

Raiffa, H. 3
Ramsey, W. 96, 139
Raphael 160
Rawls, J. 3, 71, 194
Reichenbach, R. 65
Rips, L. 36
Ritz, C. 155
Roentgen, W. C. 174
Rosen, N. 176
Russell, B. 157, 158, 235
Rutherford, E. 174, 224

Salmon, W. C. 31, 32
Savage, L. J. 178–182, 184
Sayre, K. 159
Schein 74
Schopenhauer, A. 161
Scriven, M. 31
Shakespeare, W. 130, 162, 164
Shimony, A. 177
Sibelius, P. 24
Siegel, S. 240
Skinner, B. F. 29
Smart, J. J. C. 159
Smith, C. A. B. 184

Sophocles 162
Stegmüller, W. 67, 239
Strawson, P. F. 69
Suppes, P. 175, 179, 181, 185, 186, 240, 241
Swinburne, R. 96

Talja, J. 25
Tarski, A. 181
Theatetus 205
Theorodus 204
Tranøy, K. E. 71
Tversky, A. 241

Ulm, M. 23

Wagner, C. 51, 58, 59
Weber, M. 83
Whitehead, A. N. 235
Wittgenstein, L. 65, 157
Wohlrapp, H. 67, 69

Zahar, E. G. 104
Zanotti, 177
Zeno 205
Zermelo 216

INDEX OF SUBJECTS

action 13–14, 16, 63, 68–69, 231 ff.
  explanation of 231–241
action theory 17–18
analytic-synthetic distinction 36, 47–48
analyticity 35
approximation 140–141, 143
art 160–165
attribution 70

belief 13 ff.
  change of 14–16, 18
  ethics of 13, 14, 17, 24–25
  expansion 14–15, 18
  and will 13–14

causal law 233 ff.
cause 220, 231, 232 ff.
  definition of 232
coherence 193
comment 114 ff.
  direct 116, 122, 129
  indirect 116, 124, 128
commentandum 114
commentatum 115
commenting 115
communication 111
  psychology of 124
communicative commitment 125–126,
    133
conceptual innovation 188–189
consensus (in science) 51 ff., 63, 71,
    195–197
consistency 51, 192, 193, 198
  as a goal of science 213–217
  proof 214, 215
constitution 66, 68
  v. justification 66, 70, 73–74
content (information) 18–21, 25, 196, 223
convergence
  of opinion 56 ff.

of probability assignments 55
thesis of 92, 110
cybernetics 245

decision rule
  maximax rule 5
  maximin rule 5
decision theory 239–241, 244
deliberation 9–10
deontic logic,
  see: logic
determinism 156, 172–175
dialogue
  active 128, 133
  direct 131–133
  one-sided 125
  reactive 129, 133
  types of 111, 128–133
discovery,
  context of 65
doxastic attitude 15–16, 20, 22, 24, 63

embedding 118, 120–123, 124
empirical equivalence 94–95
empirical undecidability 107–109
empiricism 64, 66, 205
  logical 171
  probabilistic 171 ff.
epistemic logic,
  see: logic
epistemic preferability 20–24
epistemic utility 18–21, 23, 71, 200
  of information (content) 18 ff.
  of truth 18 ff.
ethics 79 ff.
events
  descriptions of 234 ff.
  mental v. physical 233–234
evidence 22, 24, 25
  total 25

explanation 31 ff., 74, 220 ff.
  of action 231 ff.
  causal 231–232, 235, 242
  deductive v. inductive 30, 32–33
  and prediction 31, 220
  by reasons 231 ff., 241–245
  and understanding 29–33
explanatory power 73, 213

facts and values 83, 151–152, 157
freedom 198
  doxastic 24
  of inquiry 199
  and justice 72

genealogy of science 152, 157
grammar 124
grammatical structure 111, 124

hedonism 178
hidden variables 175
hypothetico-deductive method 222, 225

identity of indiscernibles 100
identity theory 159, 233, 244
ideology of science 191, 193, 199–200
inconsistency 84, 200
independence of irrelevant alternatives
  183
indeterminacy 107, 140, 146
  of translation 91, 238, 244
induction 73
inductive generalization
  v. physical law 154–156
information
  complex 112, 118
  new 127
  subsidiary v. main 120
  system of 113
  see also: pragmatic information
information content 18 ff.
  measure of 19
institutionalization of norms 197–198
interpretation 137 ff.
  empirical 140
  intended 145, 148
  of o-terms 148

of a predicate 140
of theoretical terms 144–146
intersubjectivity 195
invariance principles 154, 223–224

justice 71, 72, 194, 195
justification 65–66, 68, 70–71, 73, 192
  context of 65, 70
  in morality and science 81
just noticeable difference (j.n.d.) 141

language
  empirical 137 f.
  extensional 137
  interpretation of 138–140
logic
  deontic 16–18, 196, 198, 227, 228
  epistemic 22
  of inquiry 65
  many-valued 215
  of norms 196

meaning
  cognitive v. affective 35
  see also: reference, sense, similarity of
    meaning
meaning postulate 139
measurement 102, 139, 140–142, 225, 226
  errors of 172, 184
  fundamental 140, 141
  inexact 142, 183–188
  insensitivity of 142
  see also: vagueness
meta-competence 63–65, 71–72, 74
methodological rules 1 ff., 212
methodology 191, 194–195, 200
  normative principles of 217 ff.
mirror theory 157, 159–163
model 137, 139, 145–146

Newtonian mechanics 101 ff.
normative science 227–228
normative system 191–195, 198
norms 63, 71–72, 193–194, 198, 227–228
  of inquiry 191 ff.
  logic of 196
  and values 194, 196–198

o-terms 138–139, 146–148
object-competence 63–65, 71–72, 74
objectivity of research 86–87, 152
observational-theoretical dichotomy
    94 ff., 138 f.
ought-can principle 227

partition 70
philosophy of science
    analytic v. constructive 67 ff.
positivism 151
pragmatic information 113–114
    illocutionary 114
    locutionary 114
preaction 68 ff., 74, 76
preference 3
    axioms on 178–179
    epistemic 20–23
probability
    assignment 51 ff.
    consensual 54–55
    and frequency 51–52
    and rational decision making 177 ff.
    subjective 52
    upper and lower 184, 186
problem solving 188–189

quantum mechanics 171, 173, 175–177

Ramsey sentence 139
randomness 172–177
rational action 177 ff.
rational consensus
    see: consensus
rationalism 64, 66–67, 203–207
rationality
    axioms of 181–183
    Bayesian theory of 171, 177 ff.
    conditions of 80
    ethical 79 ff.
    and the explanation of action 239–240
    of practical action 80–81
    of science (scientific) 1 ff., 63 ff.,
        79 ff., 192, 199
realism 91 ff., 105 f., 109–110, 175
    and the correspondence theory of
        truth 106

realist's dilemma 91, 105 ff.
reason 5–8, 231–237, 242, 244
    practical v. theoretical 82
reference 137, 138, 145–146
relevance 112–118
    communicative 112–114, 118–119, 122
    material 113, 114
relevance distribution 118, 122–124
respect 55, 56, 59, 60
rule of ties 24

science
    and humanism 151–169
    as knowledge v. research 79
    as presentation v. research 63 ff.
    value-neutrality of 84 ff.
science policy 56, 199
scientific creativity 164–165
scientific progress 1, 2, 63
scientific rationality
    see: rationality of science
scientific tradition 221 f.
semantical system 138
semantics 137 ff.
sense 137–138, 145
similarity of meaning 35 ff.
    judgements of 38–39, 42–43
    kinds of 38–39
    scaling of 39, 46–47
simplicity 96, 97, 222
social information 52, 60
stochastic laws 173
submodel 182
success frequency 52
summary 118
summation principle 53
synonymy 35 ff.
    and similarity of meaning 47–48

testability 192, 213 f.
testing 219
text
    communicative function of 111
    microstructure of 111 f.
theoretical terms 139, 144–148
theory 91 ff., 137 ff.
    applications of 140

empirical 140–141
see also: underdetermination
theory change 137 ff.
time
closed 98–99, 101, 107–108
continuous 101, 103–105
open 100, 105, 107–108
and space 101 ff.
truth 71, 158–159
as aim of science 209–213
approximate 144, 211 ff.
and consensus 71
correspondence theory of 71, 106
criterion of 156
as epistemic utility 18, 20–21, 23–24
ethical appeal of 159–160
informative 211 ff.

underdetermination (of theory by data)
91–110

utilitarianism 178
utility
consensual 56–57
epistemic
see: epistemic utility
expected 5, 178
intellectual 57
maximization of 3 ff., 9–10, 178, 200

vagueness 140–143
values 193 ff.
conflict of 195, 196, 200
verisimilitude 104

weight (assigned to an opinion) 52 ff.
average 54 f.
of a probability assignment 54–56,
58–60
of a utility assignment 56–57

PHILOSOPHICAL STUDIES SERIES
IN PHILOSOPHY

Editors:

WILFRID SELLARS, Univ. of Pittsburgh and KEITH LEHRER, Univ. of Arizona

Board of Consulting Editors:

Jonathan Bennett, Alan Gibbard, Robert Stalnaker, and Robert G. Turnbull

20. DONALD NUTE, *Topics in Conditional Logic*, 1980.
21. RISTO HILPINEN (ed.), *Rationality in Science; Studies in the Foundations of Science and Ethics*, 1980
22. GEORGES DICKER, *Perceptual Knowledge, An Analytical and Historical Study*, forthcoming.
23. JAY F. ROSENBERG, *One World and Our Knowledge of it: The Problematic of Realism in Post-Kantian Perspective*, forthcoming.